Extrusion Handbook

Extrusion Handbook

Edited by
Steve Lawrence

STATES
ACADEMIC PRESS
www.statesacademicpress.com

Published by States Academic Press,
109 South 5th Street,
Brooklyn, NY 11249, USA

ISBN: 978-1-63989-195-5

Cataloging-in-Publication Data

Extrusion handbook / edited by Steve Lawrence.
p. cm.
Includes bibliographical references and index.
ISBN 978-1-63989-195-5
1. Extrusion process. 2. Extrusion blow molding. 3. Manufacturing processes.
4. Power presses. 5. Industrial engineering. I. Lawrence, Steve.
TJ1450 .E98 2021
621.98--dc23

For information on all States Academic Press publications
visit our website at www.statesacademicpress.com

STATES
ACADEMIC PRESS

Contents

Preface ... VII

Chapter 1 Extrusion Cooking Technology: An Advance Skill for Manufacturing of Extrudate Food Products ... 1
Tiwari Ajita

Chapter 2 Physicochemical and Rheological Changes of Starch in Nixtamalization Processes: Extrusion as an Alternative to Produce Corn Flour ... 15
Carlos Martín Enríquez Castro, Patricia Isabel Torres-Chávez, Benjamín Ramírez-Wong, Ana Irene Ledezma-Osuna, Armando Quintero-Ramos, Jaime López-Cervantes and María Irene Silvas-García

Chapter 3 Design of Polymer Extrusion Dies Using Finite Element Analysis ... 35
G.N. Kouzilos, G.V. Seretis, C.G. Provatidis and D.E. Manolakos

Chapter 4 Changes in Nutritional Properties and Bioactive Compounds in Cereals During Extrusion Cooking ... 50
Cuauhtémoc Reyes Moreno, Perla C. Reyes Fernández, Edith O. Cuevas Rodríguez, Jorge Milán Carrillo and Saraid Mora Rochín

Chapter 5 Extrusion from Gear Manufacturing to Production of Cereals ... 72
Sayyad Zahid Qamar

Chapter 6 The Extrusion Cooking Process for the Development of Functional Foods ... 81
Martha G. Ruiz-Gutiérrez, Miguel Á. Sánchez-Madrigal and Armando Quintero-Ramos

Chapter 7 Indirect Extrusion: A Multifaceted Approach of Sub-surface Tubular Expansion ... 101
Tasneem Pervez, Sayyad Z. Qamar, Omar S.A. Al-Abri and Rashid Khan

Chapter 8 Stiff Vacuum Extrusion for Agglomeration of Natural and Anthropogenic Materials in Metallurgy ... 124
Ivan Kurunov and Aitber Bizhanov

Chapter 9 Extruded Aquaculture Feed ... 145
Efren Delgado and Damian Reyes-Jaquez

Chapter 10 Extrusion Processing of Ultra-High Molecular Weight Polyethylene ... 164
Haichen Zhang and Yong Liang

Chapter 11 **Application of Open-die Warm Extrusion Technique in Spur Gear Manufacturing** 179
Wei Wang and Jun Zhao

Permissions

List of Contributors

Index

Preface

The world is advancing at a fast pace like never before. Therefore, the need is to keep up with the latest developments. This book was an idea that came to fruition when the specialists in the area realized the need to coordinate together and document essential themes in the subject. That's when I was requested to be the editor. Editing this book has been an honour as it brings together diverse authors researching on different streams of the field. The book collates essential materials contributed by veterans in the area which can be utilized by students and researchers alike.

The process of creating objects of a fixed cross-sectional profile is known as extrusion. During this process, a material is pushed through a die that has the desired cross-section. Extrusion is advantageous over other manufacturing processes because it has the ability to create very complex cross-sections. It may be continuous or semi-continuous. Continuous extrusion process can produce an indefinitely long object. Whereas semi-continuous extrusion produces many pieces. Extrusion can be done with hot as well as cold materials. Some of the commonly extruded materials are ceramics, polymers, modeling clay, metals and foodstuffs. The products made after extrusion are generally known as extrudates. The various advancements in extrusion are glanced at and their applications as well as ramifications are looked at in detail in this book. From theories to research to practical applications, case studies related to all contemporary topics of relevance to this field have been included herein. This book will serve as a valuable source of reference for graduate and post graduate students.

Each chapter is a sole-standing publication that reflects each author´s interpretation. Thus, the book displays a multi-facetted picture of our current understanding of application, resources and aspects of the field. I would like to thank the contributors of this book and my family for their endless support.

Editor

1

Extrusion Cooking Technology: An Advance Skill for Manufacturing of Extrudate Food Products

Tiwari Ajita

Abstract

The snack industry is one of the fastest growing food sectors and is an important contributor within the global convenience food market. Nowadays snacks and convenience foods are also consumed regularly in India. Properly designed convenience foods can make an important contribution to nutrition in Indian societies where social changes are altering traditional patterns of food preparation. Extrusion cooking as a popular means of preparing snack foods based on cereals and plant protein foodstuff has elicited considerable interest and attention over the past 30 years. Several studies on the extrusion of cereals and pulses, using various proportions, have been conducted because blends of cereals and pulses produce protein enriched products. Based on dough's functional properties like WAI, WSI, ER and BD, the extruded products can be classified into different group as per particular application. Therefore, this chapter is dealt with the effect of extrusion processing on product parameters, and nutritional and anti-nutritional properties of extruded product.

Keywords: extrusion processing, direct expanded snack, single screw extruder, twin-screw extruder

1. Food extrusion processing

Extrusion process is an efficient continuous process, which uniquely combines several unit operations viz.: mixing, shearing, heating, pumping, forming, and sizing. Food extruders are classified thermally as forming or cooking and geometrically as single or twin screws. Single screw forming extruders are used to manufacture pasta, processed meats, and fillings. Single screw cooking extruder (SSCE) are used to produce dry and semi moist pet foods, expanded snacks, breakfast cereals, puddings, soup and drink bases, gelatinized starch and texturized vegetable proteins. Twin-screw extruder applications include most SSCE products and chocolate coatings, candies, gums, enzyme modification process, etc. [1]. A food extruder is a high temperature short time bioreactor that transforms a variety of raw material/ingredients into finished product. Extrusion processing is a continuous process. The extruded products are sterile and because of complete starch gelatinization, very digestible [2].

Extrusion cooking is used for processing of starchy as well as materials since a long time. As extrusion processing is a thermally efficient process, it offers many advantages in processing of high protein based products like soy or legumes etc. Due to high temperature short time cooking of soy-cereal blend, the antinutritional factors are effectively destroyed without damage to nutritional quality of raw material [3].

2. Equipment and processing steps in making a direct expanded product

There are different processing steps involved for production of direct expanded snacks (**Figure 1**).

2.1. Mixed raw material/blender

This usually takes the form of a ribbon blender. The mixing tool inside the vessel is in the shape of a spiral ribbon which rotates through a reduction gear and electric motor. All the dry ingredients, along with liquid ingredients such as an emulsifier, lipids, and moisture (water), are loaded in measured amounts to the blender and mixed for the required time. Since the moisture content for an expanded product is low (less than 20%), it can be added to the blender with dry ingredients. This is a batch mixing process.

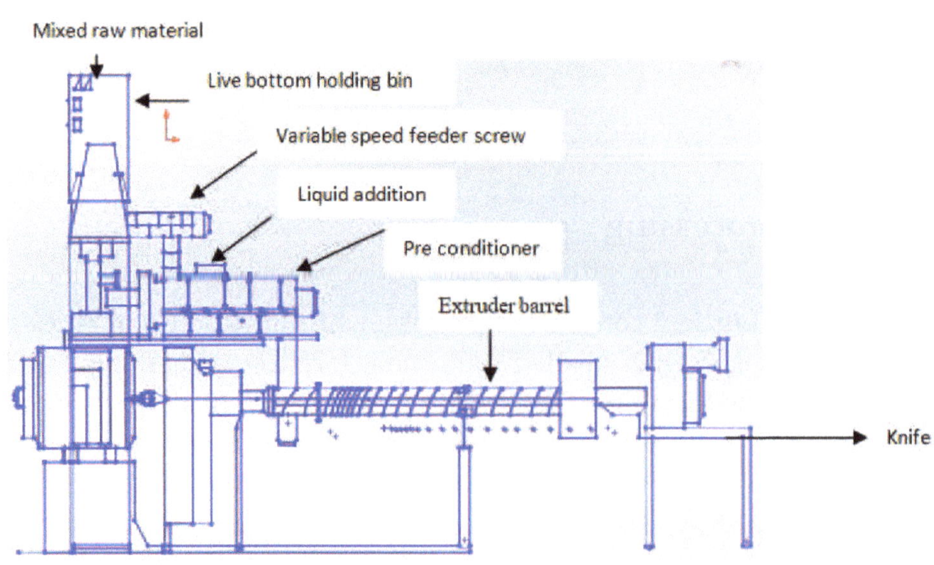

Figure 1. Layout of a typical snack food line for direct expanded products.

2.2. Variable speed feeder

This is usually in the form of an inclined screw conveyor, rotated by a geared motor, which transfers the pre blended raw-materials from the blender to the extruder hopper.

2.3. Extruder

The extruder has a hopper fitted with a horizontal auger screw run by a variable speed motor. The volumetric feeder constantly supplies a preset amount of raw-materials into the extruder inlet and over the extrusion screw running inside a grooved, electrically heated barrel. These materials are continuously moved through processing zones and forced through the die into the desired shape. Product temperature at the die exit can be as high as 190°C. Use of twin screw extruders is growing rapidly in the food industry as explained earlier. The extruder has no heating provision and the product gets sheared and temperature rises because of mechanical working of the ingredients between the plates. This extruder is almost superseded by the modern high shear cooking extruder which has versatility and immense product possibilities.

2.4. Cutters/knife

Automatic cutters are of a die-face cut variety and usually consist of a set of rotating knifes through a variable speed motor. Three dimensional cutting blades are more sophisticated and need additional knives, mounted at proper angles, to form three dimensional cut figures.

2.5. Dryer

Continuous running steel perforated belts, arranged for single or multiple passes, dry the extruded product down to 1–1.5% moisture content (wet basis). The dryer is used to produce baked collet and other products of low bulk density, whilst an additional fryer is required to produce high density products. For example, corn curls produced on a collet extruder are usually fried in a fryer to reduce the moisture level.

2.6. Coating unit

A coating unit is used to spray oil on an expanded product and to dust product with a suitable seasoning such as salt for additional mouth feel and crunch. In some units, the dryer and coating units are combined.

3. Effect of extrusion processing on product parameters

Extrusion cooking/processing of blended foods consists consideration of characteristics of starcheous and proteinaceous material i.e. gelatinization of starch and denaturation of proteinaceous material to produce quality extruded product [4]. Research carried out by different workers on effect of processing parameters on extruded snack food quality is presented below.

3.1. Expansion

The expansion is characterised on cooled and dimensionally stable products. Expansion parameters are derived both from bubble growth up until maximum size and from the ensuing contraction [5]. In a, the extrudate expansion, is a fundamentally important property during food extrusion cooking process. It is helpful in describing the product quality and also related to degree of cook. The product acceptability is based on its specific extrudate expansion. Thus, the understanding of the effects of process parameters on extrudate expansion becomes crucial for the extrusion cooking process. Several expansion theories and models have been developed to explain the characteristics of the extrudate expansion for several raw materials [6–11].

Faubion and Hoseney [12] reported that expanded volume of feed decreased with increasing amounts of proteins in the feed material, but increased with increasing starch content. In order to account for extrudates expansion upon removal from the die, longitudinal (LEI) and sectional expansion indices (SEI) proposed by Alvarez-Martinez et al. [11] were calculated.

Onwulata et al. [13] studied the effect of incorporation of whey product in extruded corn, potato and rice snacks. They concluded that the incorporation of whey protein from 0, 25 and 50%, the expansion indexes (EI) were found to be 2.4, 1.5 and 1.3, with corn flour extrudate: 2.2, 1.8 and 1.6. In case of potato flour extrudate and with rice extrudate the EI were 2.8, 2.6 and 1.8 respectively at high shear rate. Thus the effect of incorporation of whey protein with respective flours was not much as compare to flours alone. Dragnovi et al. worked fish meal, wheat gluten and soy protein blends and reported the effect of system parameters (Screw speed and barrel temperature 112–138°C) have an insignificant effect on radial expansion and in the range of 1.19–1.53. Similarly Ayse Ozer et al. [14] reported that the effect of screw speed and feed moisture on nutritious blend (Chickpea, corn, oat, corn starch, carrot powder and ground raw hazelnut) had significant effect on radial and axial expansion and were in the range of 2.36–3.08. Faubion & Hoseney [15] found that expansion of starch was greater than for wheat extrudates and decreased with increasing moisture. According to Kannadhason et al. [16], the expansion ratio of cassava and potato starch was found to decrease by 12.3 and 10.6%, respectively, with the change in net protein content from 28 to 32% wb. At higher moistures the expansion showed a maximum with respect to temperature, as reported for maize [17] and manioc starch [18]. Moraru and Kokini [9] reported that the attempt to incorporate high levels of fibre in extruded products often resulted in a compact, tough, non-crisp and undesirable texture in extrudates and reduced expansion. Falcone and Phillips [19] studied sorghum and cowpea blend and found that both temperature (175–205°C) and moisture (20.5–25%) had negative effect on expansion for most compositions. While various studies on extrusion of proteinaceous [20] and starchy [21] systems have found that puffing is directly related to temperature and inversely related to moisture. They observed that adding protein to a starchy extrusion system may interfere with expansion and also that amylopectin exerts a positive and amylose a negative influence on expansion.

Altan et al. [22] studied the effect of die temperature (140–160°C), screw speed (150–200 rpm) and pomace level (2–10%) on barley-grape pomace extrudate and found that effect of temperature had more effect. EI decreased with increasing barrel temperature and the value ranged between 0.949 and 1.747.

Ding et al. [23] studied the effect of extrusion conditions on physicochemical properties of rice based snacks and feed moisture was found to be main factor affecting the extrudate expansion. The highest expansion (3.87) was reported at 14% feed moisture, 120°C barrel temperature and screw speed at 250 rpm.

Molla [24] reported that in case of wheat extrudates, with increase in the screw speed increased from 200 to 300 rpm, initially sectional expansion index increased from 9.15 to 10.54, and then decreased to near the initial expansion. Whereas for corn extrudates no significant evolution occurred between 200 and 300 rpm, however a significant drop (39%) below the initial expansion was recorded for a speed of 500 rpm. Otherwise, the increase in screw speed induced a significant rise in the longitudinal expansion of extrudates for the two types of flour. The LEI for wheat and corn extrudates displayed an overall increase of 43 and 46%, respectively, for an increase in speed from 200 to 500 rpm.

3.2. Bulk density

The extrudate density was mainly affected by feed moisture. Screw speed and temperature also have significant effects on the density of extrudate. Increased feed moisture also promotes a sharp increase in extrudate density. However, increased screw speed and barrel temperature caused a slight decrease in the density of extrudate. Ding et al. [23] studied the effect of extrusion conditions on physicochemical properties of rice based snacks and feed moisture was found to be main factor affecting the extrudate expansion. The lowest bulk density (0.1 g/cm^3) was reported at lowest feed moisture (14%) and highest barrel temperature (140°C).

Altan et al. [22] studied the effect of die temperature (140–160°C), screw speed (150–200 rpm) and pomace level (2–10%) on barley-grape pomace extrudate and found that both (pomace level and barrel temperature) had significant effect on bulk density. The bulk density of extrudates was ranged between 0.325 and 1.18 g/cm^3. The increase in temperature from 140 to 150°C decreased the bulk density from 0.85 to 0.25 g/cm^3, whereas increase of pomace level increased the bulk density from 0.325 to 0.95 g/cm^3. The highest BD 1.18 g/cm^3 was found at 140°C and 10% pomace.

Feed moisture has been found to be the main factor affecting extrudate density and expansion [15, 25–27]. With an increased feed moisture content during extrusion due to plasticization of the melt may reduce the elasticity of the dough. This promote to in reduce SME and therefore reduced gelatinization, decreasing the expansion and increasing the density of extrudate.

It was observed that extrudate density is inversely affected with an increase in screw speed. Increase in screw speed lowers the melt viscosity of the mix increasing the elasticity of the dough, resulting in a reduction in the density of the extrudate [25]. An increase in the barrel temperature will increase the degree of superheating of water in the extruder encouraging bubble formation and also a decrease in melt viscosity [25] leading to reduced density. Similar results have been observed by Mercier and Feillet [28]. The bulk density of extrudate increased with decreasing expansion ratio. Expansion and bulk density are also related to starch gelatinization [29]. According to these authors, an increase in gelatinization increased expansion and decreased bulk density.

3.3. Water absorption index (WAI) and water solubility index (WSI)

The WAI measures the amount of water absorbed by starch and can be used as an index of starch gelatinization [23, 30, 31]. WSI, often used as an indicator of degradation of molecular components [32], measures the amount of soluble components released from the starch after extrusion. When extruded products mixed with water, this mixture will often swell. Out of that a portion of material will become soluble. Water solubility and absorption are often important in predicting the extruded material behaviour if further processed [33].

Water absorption index indicates the amount of water immobilised by the extrudate, while water solubility indicates the amount of small molecules solubilised in water so process molecular damage. Anderson et al. [30] recorded a method to estimate the amount of material that can be extracted by water from an extruded product. The materials which are soluble include gelatinized starch; undenatured globular proteins, inorganic ions and small sugars [33]. The WSI increased significantly when screw speed increased from 200 to 300 rpm for wheat extrudates and from 300 to 500 rpm for corn extrudates [34]. The WSI is indeed related to the degree of starch transformation. The unprocessed flours exhibited values of WSI less high than those of final products (for corn flours). Consequently, the WSI increased because starch granules were then more soluble in water [35].

WAI increased with extrusion temperature and feed moisture content for corn and corn-lentil extrudates [36]. The WAI measures the amount of water absorbed by starch and can be used as an index of gelatinization, since native starch does not absorb water at room temperature [30, 31, 37]. Extrusion temperature and moisture content are known to affect gelatinization during extrusion, and consequently the WAI. In high moisture soy meat analog, WAI increased with increase in extrusion temperature and feed moisture [38, 39]. Similar results were reported for corn starch extrudates, bean and chickpea extrudates [40, 41].

Furthermore, dextrinization is well known as the predominant mechanism of starch degradation during low moisture extrusion. Therefore, the decreasing trend of WSI with feed moisture content is expected and in agreement with previous reports [23, 42].

3.4. Product moisture

Product moisture was found to be directly related to feed moisture and inversely related to extrusion temperature [43]. After drying at 60°C for 12 h of starch-PDPF extrudates, the moisture content was found to be very low nearly 0.5% which expected to yield products with a high degree of crispness. After drying of extrudates the Water activity dropped down from 0.1 to 0.33 that would be advantageous with regard to the stability of the extrudate against microbial growth.

Moisture is having significant on product quality attributes such as expansion and degree of cook (absorption and solubility indices) [32]. It is necessary to adjust the water content carefully to result in expansion with whey incorporated products. Increased structural binding of water may have reduced moisture available for flash-off and consequently reduced expansion [44].

3.5. Specific mechanical energy (SME)

The specific mechanical energy (SME) is responsible for fragmentation of starch molecules [21, 45, 46]. Amylopectin molecules are broken mainly at the α–1:6 bonds due to the applied shear forces. This phenomenon was attributed to the decrease in the viscosity with the increase in water content [21, 45]. The degradation products are macromolecules in the range of 50,000–200,000 MW [46, 47].

An additional effect of SME on starch is the gelatinization process that takes place during extrusion [21, 45, 46]. The degree of gelatinization would be with the higher value of SME. In contrast to the effect of water on macromolecule fragmentation, gelatinization of starch is more intense at higher water content.

Mercier et al. [4] reviewed that SME input also depends on the exact composition of the product being extruded and increases with starch content. A general result is that SME increases when water content decreases in both single screw and twin screw [44].

3.6. Effect of extrusion on nutritional constituents

3.6.1. Proteins

Proteins are a group of highly complex organic compounds that are made up of a sequence of amino acids. Protein nutritional value is dependent on the quantity, digestibility and availability of essential amino acids [40].

Several changes occur during extrusion of which denaturation is undoubtedly the most important. Extrusion may improve protein digestibility by denaturating proteins and exposing enzyme-accessible sites [37, 48, 49]. Enzymes and enzyme inhibitors generally lose activity due to denaturation. Protein digestibility value is higher for non-extruded products. The possible cause might be the denaturation of proteins and inactivation of anti-nutritional factors that impair digestion. The extensive studies have been done and reported on the effects of extrusion on protein nutrition especially for animal feeds and for human weaning foods [50]. The extrusion operations have very little effect on the protein denaturation [51]. Maillard reactions occur during extrusion particularly at high barrel temperature, low moisture, and high shear. All processing variables have different effects on protein digestibility. High shear extrusion conditions in particular promote denaturation [52], although mass temperature and moisture are also important factors. In a model system of wheat starch, glucose and lysine, low pH favours Millard reactions, as measured by increased colour [53].

Cooking extruders for processing high-protein materials into palatable foods is very common today. Many new applications have been developed for protein extrusion during the past decade. Improvements in functional characteristics of proteins may be achieved through modification of temperature, screw speed, moisture content, and other extrusion parameters.

3.6.2. Vitamins

During extrusion process due to vast deviation of chemical structure and composition of vitamins, there is variable change in variety. The extent of degradation depends on different process parameters and storage conditions such as moisture, temperature, light, oxygen, time and pH [54, 55].

Among the fat-soluble vitamins, vitamins D and K are fairly stable [56]. In food extrusion process the thermal degradation is the major factor contributing to β-carotene losses [57].

Pham and Del Rosario [58] and Guzman-Tello and Cheftel [59] studied the effects of high temperature, short-time extrusion cooking on vitamin stability and developed different mathematical models.

In extrusion cooking there is inverse relation between the retention of vitamins and temperature, screw speed and specific energy input, whereas direct relation with moisture, feed rate and die diameter.

3.6.3. Iron and zinc

During extrusion cooking the mineral contents are generally retained well. During single screw extrusion of potato flakes with increase in barrel temperature there is increase in, iron content [60]. Total iron increased by as much as 38% due to extrusion [61]. On the other hand, after twin screw extrusion cornmeal (having low dietary fibre content) had no changes in total, elemental, or soluble iron [55].

Utilisation of iron and zinc from wheat bran and wheat in adult human volunteers was not affected by extrusion [62]. Low-shear extrusion retained dialysable iron in navy beans, lentils, chickpeas and cowpeas better than did high-shear extrusion [63]. Weaning food blends of pearl millet, cowpea and peanut had greater iron availability and protein digestibility compared to similar foods processed by roasting [64].

3.7. Antinutrient factors

3.7.1. Antinutrients

Extrusion cooking also improves the nutritional quality of foods by destroying many natural toxins and antinutrients (**Table 1**). A dilemma exists as to whether it is desirable to remove these compounds. Enzyme inhibitors, hormone-like compounds, saponins and other compounds could impair growth and development in children, but these same compounds may offer protection against chronic diseases in adults.

3.7.2. Phenolic compounds

Extrusion of soy protein concentrate and a mixture of 80:20 of cornmeal and soy protein concentrate (80:20) did not result in changes in total isoflavone content [65]. In potato peels

Compound	Foods	Factors favouring reduction
Allergens	Peanuts, soy	Increased shear; added starch
Glucosinolates	Canola	Added ammonia
Glycoalkaloids	Potato	Added thiamine
Gossypol	Cottonseed	Higher feed moisture
Mycotoxins	Grains	Increased mixing, lower temperatures; added amine sources
Protease inhibitors	Legumes, potato	Higher extrusion temperature

Table 1. Antinutrients and toxins affected by extrusion cooking.

produced by steam peeling during extrusion the total free phenolics, primarily chlorogenic acid, decreased significantly [55]. More phenolics were retained with higher barrel temperature and feed moisture. It might be possible that lost phenolics reacted with themselves or with other compounds to form larger insoluble materials. The total antioxidant activity value of samples decreased with an increase in screw speed and decrease in moisture content, while total phenolic values had insignificant (95% confidence interval) changes after extrusion. In a model breakfast cereal, containing cornmeal and sucrose, anthocyanin pigments were degraded at higher levels of added ascorbic acid, and total anthocyanins significantly decreased by extrusion [66].

Many opportunities exist for product development research in extrusion. Although several studies have been conducted on determining the effect of raw material combination and process parameters on physic-chemical characteristics of direct expanded snacks as well as their storage studies. Very little has been published on the effects of extrusion on phytochemicals and other healthful food components, in part due to the need for identification of active principles and suitable analytical procedures. Evaluations of nutrient retention by either high-moisture extrusion or by supercritical fluid extrusion have yet to be published. Improved understanding of scale-up issues in extrusion is necessary for valid interpretation of studies conducted using laboratory-scale and pilot plant extruders. Long-term animal and feeding studies are tedious and costly, yet essential for demonstrating safety and efficacy of extruded foods.

4. Conclusions

Extrusion cooking is one of the most important food processing technologies which have been used for the production of breakfast cereals, ready to eat snack foods and other textured foods. Now a days extrusion cooking is a widely used technology in the agri-food processing industry. Therefore adopting such skill can provide the good opportunity to the snack industries for developing large variety of food snacks. It is a popular unit operation for producing a variety of food products with numerous ingredients requiring a wide range of processing conditions and includes starch, protein, lipids, water and additives. There are ambiguous effects of extrusion cooking on nutritional quality of expanded snacks. Because of its beneficial effects such as destruction of antinutritional factors, increased soluble dietary fibres, reduction of lipid oxidation and contaminating microorganisms, it plays an important role in the production of a wide variety of foods and ingredients. Extrusion cooking being a complex multivariate process, to maintain the product quality requires careful control on the process. Severe extrusion conditions and improper formulation which are not suitable for process may cause nutritional destruction in the hot-screw segments. Generally, to maintain high nutritional quality, high extrusion temperature (≥200°C) and low moisture content (≤15%) of the feed should be avoided. There are many areas that require further research regarding extrusion and nutrition. Future research may be focussed on the relationships between compositional changes on product quality-both nutritional and sensory aspects, and the effects of interactions between complex extruder conditions on nutrient retention.

Author details

Tiwari Ajita

Address all correspondence to: ajitatiwari@gmail.com

Department of Agricultural Engineering, Assam University, Silchar, India

References

[1] Harper JM. Food extrusion. CRC Critical Reviews in Food Science and Nutrition. 1978; **11**(2):155-215

[2] Seib PA. An Introduction to Food Extrusion. Manhattan: Kansas State University; 1976

[3] Bordoloi R, Ganguly S. Extrusion technique in food processing and a review on its various technological parameters. Indian Journal of Science, Research and Technology. 2014;**2**(1):1-3

[4] Mercier C, Linko P, Harper JM. Extrusion Cooking. St. Paul., Minnesota: American Assoc. Cereal Chemists; 1989

[5] Della Valle G, Colonna P, Patria A. Influence of amylose content on the viscous behavior of low hydrated molten starches. Journal of Rheology. 1996;**40**(3):347-362

[6] Chevanan N, Muthukumarappan K, Rosentrater KA, Julson JL. Effect of die dimensions on extrusion processing parameters and properties of DDGS-based aquaculture feed. Cereal Chemistry. 2007a;**84**(4):389-398

[7] Chevanan N, Rosentrater KA, Muthukumarappan KA. Twin screw extrusion processing of feed blends containing distillers grains with soluble (DDGS). Cereal Chemistry. 2007b;**84**(5):428-436. http://dx.doi.org/10.1094/CCHEM-84-5-0428

[8] Chevanan N, Rosentrater KA, Muthukumarappan KA. Effect of DDGS, moisture content, and screw speed on the physical properties of extrudates in single screw extrusion. Cereal Chemistry. 2007c;**90**(3):530-535. http://dx.doi.org/10.1094/CCHEM-85-2-0132

[9] Moraru CI, Kokini JL. Nucleation and expansion during extrusion and microwave heating of cereal foods. Comprehensive Reviews in Food Science and Food Safety. 2003; **2**(4):120-138

[10] Shukla CY, Muthukumarappan K, Julson JL. Effect of single screw extruder die temperature, amount of distillers dried grains with solubles (DDGS) and initial moisture content on extrudates. Cereal Chemistry. 2005;**82**(1):34-37

[11] Alvarez-Martinez L, Kondury KP, Harper JM. A general model for expansion of extruded products. Journal of Food Science. 1988;**53**(2):609-615

[12] Faubion JM, Hoseney RC. High-temperature short-time extrusion cooking of wheat starch and flour. II. Effect of protein and lipid on extrudate properties. Cereal Chemistry. 1982;**59**(6):533-537

[13] Onwulata CI, Smith PW, Konstance RP, Holsinger VH. Incorporation of whey products in extruded corn, potato or rice snacks. Food Research International. 2001;**34**(8):679-687

[14] Ayse Ozer E, Emine NH, Guzel S, Paul A, Şenol İ. Effect of extrusion process on the antioxidant activity and total phenolics in a nutritious snack food. International Journal of Food Science & Technology. 2011;**41**(3):289-293

[15] Faubion JM, Hoseney RC. High temperature short time extrusion cooking of wheat starch and flour. I. Effect of moisture and flour type on extrudate properties. Cereal Chemistry. 1982;**59**(6):529-533

[16] Kannadhason S, Muthukumarappan K, Rosentrater KA. Effect of ingredient and extrusion parameters on aquafeeds containing DDGS and tapioca starch. Journal of Aquaculture Feed Science & Nutrition. 2009;**1**(1):6-21

[17] Mercier C. Structure and digestibility alterations of cereal starches by twin-screw extrusion – cooking. In: Food Process Engineering. Vol. I. London: Applied Science Publisher Ltd.; 1979. pp. 795-807

[18] Mercier C, Charbonniere R, Grebaut J, Dela Gueriviere JF. Formation of amylase-lipid complexes by twin screw extrusion cooking of manioc starch. Cereal Chemistry. 1980;**57**(1):4-9

[19] Falcone RG, Phillips RD. Effects of feed composition, feed moisture, and barrel temperature on the physical and rheological properties of snack-like products prepared from cowpea and sorghum flours by extrusion. Journal of Food Science. 1988;**53**(5): 1464-1469

[20] Lawton JW, Davis AB, Behnke KC. High temperature short-time extrusion of wheat gluten and bran-like fraction. Cereal Chemistry. 1985;**62**(4):267-269

[21] Gomez MH, Aguilera JM. A physicochemical model for extrusion of corn starch. Journal of Food Science. 1984;**49**(1):40-43

[22] Altan A, McCarthy KL, Maskan M. Evaluation of snack foods from barley-tomato pomace blends by extrusion processing. Journal of Food Engineering. 2008;**84**(2):231-242

[23] Ding Q, Ainsworth P, Tucker G, Marson H. The effect of extrusion conditions on the physicochemical properties and sensory characteristics of rice-based expanded snacks. Journal of Food Engineering. 2005;**66**(3):283-289

[24] Molla A. Effect of Extrusion Operating Conditions on Aflatoxin Reduction and Product Characteristics of Corn-Peanut Flakes. 2010. http://etd.aau.edu.et/dspace/bitstream [Accessed: 10-03-12]

[25] Fletcher SI, Richmond P, Smith AC. An experimental study of twin screw extrusion cooking of maize grits. Journal of Food Engineering. 1985;**4**(4):291-312

[26] Ilo S, Liu Y, Berghofer E. Extrusion cooking of rice flour and amaranth blends. Food Science and Technology. 1999;**32**(2):79-88

[27] Launay B, Lisch JM. Twin-screw extrusion cooking of starches: Flow behavior of starch pastes, expansion and mechanical properties of extrudates. In: Jowitt R, editor. Extrusion Cooking Technology. London: Applied Science Pub., Ltd.; 1983. pp. 159-160

[28] Mercier C, Feillet P. Modification of carbohydrate components by extrusion cooking of cereal products. Cereal Chemistry. 1975;**52**(3):283-297

[29] Case SE, Hanna MA, Schwartz SJ. Effect of starch gelatinization on physical properties of extruded wheat and corn-based products. Cereal Chemistry. 1992;**69**(4):401-404

[30] Anderson RA, Conway HF, Pfeifer VF, Griffin EL. Roll and extrusion cooking of sorghum grits. Cereal Science Today. 1969;**14**(11):372-381

[31] Ding Q, Ainsworth P, Tucker G, Marson H. The effect of extrusion conditions on the functional and physical properties of wheat based expanded snacks. Journal of Food Engineering. 2006;**73**(2):142-148

[32] Kirby AR, Ollett AL, Parker R, Smith AC. An experimental study of screw configuration effects in the twin-screw extrusion-cooking of maize grits. Journal of Food Engineering. 1988;**8**(4):247-272

[33] Hill SE, Norton C. Estimation on extruded products: Rapid detection methods for water solubility and absorption measurements. In: Extruded Cereal Products: Their Creation and Evaluation. Symposium, Nottingham; 1995;28-29

[34] Mezreb K, Goullieux A, Ralainirina R, Queneudec M. Application of image analysis to measure screw speed influence on physical properties of corn and wheat extrudate. Journal of Food Engineering. 2003;**57**(2):145-152

[35] Smith AC. Studies on the physical structure of starch-based materials in the extrusion cooking process. In: Kokini JL, Ho CT, Karwe MV, editors. Food Extrusion Science and Technology. New York: Marcel Dekker, Inc.; 1992. pp. 570-618

[36] Lazou A, Krokida M. Structural and textural characterization of corn-lentil extruded snacks. Journal of Food Engineering. 2010;**100**(3):392-408

[37] Colonna P, Tayeb J, Mercier C. Extrusion cooking of starch and starchy products. In: Mercier C, Linko P, Harper JM, editors. Extrusion Cooking. St. Paul: American Association of Cereal Chemists; 1989. pp. 247-320

[38] Badrie N, Mellowes WA. Texture and microstructure of cassava (*Manihot esculenta* Crantz) flour extrudate. Journal of Food Science. 1991;**56**(5):1319-1322

[39] Lin S, Huff HE, Hsieh F. Texture and chemical characteristics of soy protein meat analog extruded at high moisture. Journal of Food Science. 2000;**65**(2):264-269

[40] Singh B, Sekhon KS, Singh N. Effects of moisture, temperature and level of pea grits on extrusion behaviour and product characteristics of rice. Food Chemistry. 2007; **100**(1):198-202

[41] Meng X, Threinen D, Hansen M, Driedger D. Effects of extrusion conditions on system parameters and physical properties of a chickpea flour-based snack. Food Research International. 2010;**43**(2):650-658

[42] Hernandez-Diaz JR, Quintero-Ramos A, Barnard J, Balandran-Quintana RR. Functional properties of extrudates prepared with blends of wheat flour/pinto bean meal with added wheat bran. Food Science and Technology International. 2007;**13**(4):301-308

[43] Maurice TJ, Stanley DW. Texture-structure relationships in textured soy protein IV. Influence of process variables on extrusion texturization. Canadian Institute of Food Science and Technology Journal. 1978;**11**(1):1-5

[44] Bhattarcharya M, Hanna MA. Influence of process and product variable on extrusion energy and pressure requirements. Journal of Food Engineering. 1987;**6**(2):153-163

[45] Gomez MH, Aguilera JM. Changes in the starch fraction during extrusion-cooking of corn. Journal of Food Science. 1983;**48**(2):378-381

[46] Van Lengerich. Influence of extrusion processing on on-line rheological behaviour, structure, and function of wheat starch. In: Thermal Processing and Quality of Foods. London: Elsevier Applied Science Publication; 1990. pp. 421-471

[47] Politz ML, Timpa JD, Wasserman BP. Quantitative measurement of extrusion-induced starch fragmentation products in maize flour using nonaqueous automated gel-permeation chromatography. Cereal Chemistry. 1994;**71**(6):532-536

[48] Van Zuilichem DJ, Van Roekel GJ, Stolp W, Van't Riet K. Modeling of enzymatic conversion of cracked corn by twin screw extrusion cooking. Journal of Food Engineering. 1990;**12**(1):13-38

[49] Wen LF, Rodis P, Wasserman BP. Starch fragmentation and protein insolubilization during twin-screw extrusion of corn meal. Cereal Chemistry. 1990;**67**(3):268-275

[50] Alonso R, Rubio LA, Muzquiz M, Marzo F. The effect of extrusion cooking on mineral bioavailability in pea and kidney bean seed meals. Animal Feed Science and Technology. 2001;**94**(1-2):1-13

[51] Areas JAG. Extrusion of proteins: Critical reviews. Food Science and Nutrition. 1992;**32**:365-392

[52] Della-Valle G, Colonna P, Patria A. Influence of amylose content on the viscous behavior of low hydrated molten starches. Journal of Rheology. 1994;**40**(3):347-362

[53] Bates L, Ames JM, Macfougall DB. The use of a reaction cell to model the development and control of colour in extrusion cooked foods. Lebensmittel Wissenschaft und Technologie. 1994;**27**(4):375-379

[54] Bjorck I, Asp NG. The effects of extrusion cooking on nutritional value, a literature review. Journal of Food Engineering. 1983;**2**(4):281-308

[55] Camire ME. Chemical changes during extrusion cooking: Recent advances. In: Shahidi F, Ho C-T, Van Chuyen N, editors. Process-Induced Chemical Changes in Foods. Vol. 434. New York: Plenum Press Div., Plenum Publishing Corp; 1998. pp. 109-121

[56] Killeit U. Vitamin retention in extrusion cooking. Food Chemistry. 1994;**42**(9):149-155

[57] Guzman-Tello R, Cheftel JC. Colour loss during extrusion cooking of beta-carotene—wheat flour mixes as indicator of the intensity of thermal and oxidative processing. Journal of Food Science & Technology. 1990;**25**(4):420-430

[58] Pham CB, Del Rosario. Studies on the development of textured vegetable products by extrusion process I. Effect of processing variable on protein properties. Journal of Food Technology. 1986;**19**:535-547

[59] Guzman-Tello R, Cheftel JC. Thiamine destruction during extrusion cooking as an indicator of the intensity of thermal processing. Journal of Food Science & Technology. 1987;**22**(5):549-562

[60] Maga JA, Sizer CE. Ascorbic acid and thiamine retention during extrusion of potato flakes. Lebensmittel Wissenschaft und Technologie. 1978;**11**(4):192-194

[61] Camire ME, Zhao J, Violette DA. In vitro binding of bile acids by extruded potato peels. Food Chemistry. 1993;**41**(12):2391-2394

[62] Fairweather-Tait SJ, Portwood DE, Symss LL, Eagles J, Minski MJ. Iron and zinc absorption in human subjects from a mixed meal of extruded and non-extruded wheat bran and flour. American Journal of Clinical Nutrition. 1989;**49**(1):151-155

[63] Ummadi P, Chenoweth WL, Vebersax MA. The influence of extrusion processing on iron dialyzability, phytates and tannins in legumes. Journal of Food Processing and Preservation. 1995;**19**(2):119-131

[64] Cisse D, Guiro AT, Diaham B, Souane M, Doumbouya NT, Wade S. Effect of food processing on iron availability of African pearl millet weaning foods. International Journal of Food Science & Nutrition. 1998;**49**(5):375-381

[65] Mahungu SM, Diaz-Mercado SLJ, Schwenk M, Singletary K, Faller J. Stability of isoflavones during extrusion processing of corn/soy mixture. Journal of Agricultural and Food Chemistry. 1999;**47**(1):279-284

[66] Camire ME. Bilberries and blueberries as functional foods and pharmaceuticals. In: Mazza G, Oomah DB, editors. Functional Foods: Herbs, Botanicals and Teas. Lancaster, PA: Technomic Press; 2000. pp. 289-319

2

Physicochemical and Rheological Changes of Starch in Nixtamalization Processes: Extrusion as an Alternative to Produce Corn Flour

Carlos Martín Enríquez Castro,
Patricia Isabel Torres-Chávez,
Benjamín Ramírez-Wong,
Ana Irene Ledezma-Osuna,
Armando Quintero-Ramos,
Jaime López-Cervantes and
María Irene Silvas-García

Abstract

Corn tortilla is a food consumed mainly in México and Central America. It provides 50% of total calories ingestion and is a good source of fiber. Tortilla is produced by the nixtamalization process using corn, water and lime. It has been produced by alternative processes as extrusion, reducing cooking liquor, and increasing dietary fiber. The aim of this book chapter is to describe the changes in corn starch by different nixtamalization processes, also are presented the advantages and disadvantages of both processes, encouraging some aspects of producing corn flour by extrusion. The extrusion is a technology that is dependent of process variables and is reflected on quality of end product. Several factors are involved, as feed moisture and temperature, and they have a direct impact on corn starch physicochemical, textural, and rheological properties.

Keywords: starch, nixtamalization, extrusion, corn flour, tortilla

1. Introduction

In México, the main use of maize is the transformation to nixtamal (cooked corn) then into masa (corn dough), tortilla (main product), tamal, or snacks [1, 2]. Tortilla has been considered a good source of fiber and calcium [3] and is staple food in this country.

There are several nixtamalization processes to make those products. The most common is the traditional nixtamalization process (TNP), an ancient process where corn is cooked with water and lime, steeped in alkaline liquor, and washed to remove cooking water. Clean and cooked corn is named nixtamal, which is ground to produce corn masa, then tortilla [4]. Fresh masa is produced in small establishments and also at strict and highly efficient facilities [5]. Fresh corn masa is also dried and grounded to produce corn flour [6]. On the other hand, nixtamalized corn flour (NCF) is produced at industrial scale reducing water, lime, and steeping time, producing nixtamal, which is dried and exposed to successive stages of milling. Fine particle size in the flour is required, getting an easily hydrated product [7].

In both processes, the discharge of cooking liquor (named nejayote with alkaline pH) has an ecological impact [8]. Consequently, economic and commercial implications are involved because high concentrations of soluble solids are discharged in the effluents (2–11%) [9].

Extrusion is a continuous process in which raw matter is transformed in dough because of a combination of shear, feed moisture, and temperature [10]. It has been reported that extrusion cooking increases nutritional content, hygiene, physicochemical and sensorial characteristics of end product, inactivate anti-nutritional agents, and enzymes from raw matter used to elaborate extruded products [11]. Feed moisture content, particle size distribution (PSD), screw configuration, speed, die size, and heat input are important in the quality of end product. Using correctly all these factors, an accelerated process of gelatinization and fragmentation in starch granule can be prevented [12]. Extrusion normally increases the generation of resistant starch [13], and several times melting and fragmentation reactions are initiated [14]. The feed moisture in extrusion is low, water consumption diminishes, and production of alkaline effluent is absent [8].

Corn masa texture is an important factor to consider when tortilla is produced [15]. Tortilla production depends on changes in corn starch of the nixtamalization process, since raw corn is processed through several stages until production of tortilla is done [16, 17]. On each step of the process, changes in starch are in different degree (less or more damage). Good quality of tortilla and others nixtamalized products is synonym of an adequate cooking. Quality is reflected as good cohesiveness and adhesiveness in masa, which means a good performance in the forming rollers. A good process control is performed when the tortilla is formed adequately without being so sticky and gets stuck in the forming rollers. Undercooked masa has poor adhesiveness and causes troubles in the forming rollers [18]. The aim of this review is to describe physicochemical and rheological changes in starch during the traditional nixtamalization process and to compare to those occurring during nixtamalization using the extrusion process. The advances and benefits of producing instant corn flour using the extrusion are also described.

2. Starch

2.1. General characteristics

Starch is found in seeds, roots and tubers, stems, leaves, fruits and even pollen, being organized as discrete particles named granules [19]. Starch granule is insoluble in cold water and is present as spheres, ellipsoids, circles, and other irregular shapes. Dimensions of starch granule range from 0.1 to 200 μm [20], and the solubility of starch is increased when warm water is used. Starch is a complex carbohydrate composed of 20–25% amylose and 75–80% amylopectin [21], being the main component of maize kernel.

Amylose is almost linear glucose molecule linked by α-1, 4 D-glucopyranosil linkages; with a molecular weight (MW) of 105×10^6 g/mol [22]. Amylopectin integrates anhydroglucose units in a highly structured architecture made of short α-1,4 glucan chains (95%) and α-1,6 linkages (5%) with a MW of 107×10^8 g/mol. Amylopectin defines most of the chemical and physical properties of starch from different sources [23]. Starch properties have been attributed to differences in amylopectin structure. These properties include granule swelling (onset of viscosity), peak viscosity, peak temperature, shear thinning during pasting, and gel firmness during storage [24]. They are related with the quality of end product.

2.2. Functional properties of starch

Functionality is concerned with rheological and structuring properties obtained after cooking. Functional properties of starch are associated with molecular level effects (gelatinization temperature, solubilized starch, and retrogradation) and granule level behavior (swelling, solubility index, and rheology of swollen granules) [20, 25]. The functionality of starch granule changes when it is heated in presence of water. Gelatinization changes are manifested as a loss in crystallinity, granular swelling, and increase in solubility [25, 26].

The crystallinity of starch is attributed to linear short chains presented in the amylopectin molecules. It is represented as a three-dimensional crystalline structure shown by X-ray diffraction (X-RD) patterns [23]. Native starch granules have a crystallinity level ranging from 15 to 45% [20] present in an amorphous state and showing the characteristic patterns of cereals (type A) [27]. When the gelatinization process begins in corn starch, which takes place between 70 and 75°C, the crystalline and organized structure of starch is transformed into an amorph state [20], and crystallinity is lost [28].

The leaching of amylose in starch granule is provoked by several factors. Some of the most representative ones are an increase in temperature, presence of other solutes, type and concentration of the starch, and the agitation force applied during heating [27, 29]. Associated with these changes, the retrogradation begins as a kind of internal restructuring in starch, creating a more compact and solid molecule [30, 31]. Retrogradation is dependent on macromolecular structure (chain configuration, ramifications, and distribution of molecular weight) and the botanical source [32].

Starch granules are physically and chemically inert and are not very digestible in the human body. To change them into functional products, they are heated in excess of water and eventually pass from a semicrystalline and relatively indigestible form to an easily digestible amorphous form [30]. As the structure begins to weaken, the granules soak water and swell. Since not all the granules are gelatinized simultaneously, different degrees of swelling and structural disorganization may exist. This process is named annealing and takes place during soaking for a certain period of time at sub gelatinization temperatures, whereby the starch undergoes reorganization in a more ordered structure [31].

3. Nixtamalization processes

3.1. Traditional nixtamalization process (TNP)

The nixtamalization process is performed actually in many countries, especially in México and Central America [4]. The emerging tortilla industry in the United States has the fastest growing segment of the baking industry in the U.S. market, estimating that Americans in year 2000 consumed 85 million tortillas according to Tortilla Industry Association (TIA) [32].

Functional parameters evaluated in nixtamalization process give us an idea of changes taking place in main components of starch, responsible for rheological and textural properties in nixtamalized corn products [33, 34]. Corn starch crystallinity in the nixtamalization process evaluated with X-RD technique usually presents similar defined peaks that correspond to the interplanar spacing values "d" of 5.86, 5.19, 4.90, 4.46, and 3.87Å [16, 33]. During the TNP, the maize loses part of the typical crystalline structure denoting the formation of amylose-lipid complexes [35]. Studies of changes in crystallinity patterns of corn starch using X-ray diffraction caused by the nixtamalization process have been cited in the literature [16, 36–38]. Other investigators [16, 33, 39] have reported loss of Maltese cross of starch during the nixtamalization process of tortilla. Loss of birefringence indicates a loss of molecular order and general molecular reorganization within starch granules [38].

The starch granule swelling begins with the application of heat, changing its structure. Water is introduced within the granule, and H-bonds between water molecules and polar residues of glucose units are transformed [40]. The high amount of polar groups accelerates the water absorption, and the starch granule collapses. After cooking, linkage of amylose and amilopectin happens and then aggregates outside of the internal structure and forms a gel [16, 22].

Structural damage in aleurone layer and some pericarp *strata* of corn were observed using scanning electronic microscopy (SEM) during alkaline cooking to produce nixtamal [40]. Gómez et al. [39] evaluated the microstructural changes of starch during all the stages of the nixtamalization process, and they concluded that the greatest damage in starch was produced during baking of tortilla because of the use of high temperatures (190°C) and also when tortilla chips were produced in the deep frying process.

The thermal characterization is important to define the cooking variables of maize and its products. The nixtamalization process requires maize with low temperature and enthalpy of gelatinization for tortilla production [41]. These values for maize starch are between 67 and 69°C for temperature and 8–16 J/g of starch for enthalpy [42]. Other gelatinization temperature ranges of nixtamalized processed products have been monitored between 70 and 80°C [33, 43], defining their structural and textural characteristics [42]. An extended cooking time and steeping involve more gelatinized starch and lower enthalpy values [5] increasing the gelatinization temperature and producing a more reorganized starch molecule. This parameter also involves a higher degree of crystallinity retrogradation, shortening, and syneresis (water is squeezed out of the granule) of overcooked starch [39]. All these changes are related with corn masa yield [44].

3.1.1. Stages in TNP

3.1.1.1. Cooking and steeping

Cooking of corn is used to hydrate corn kernel, tender the pericarp, denaturalize proteins, and develop partial gelatinization of starch because of absorption of lime in germ [38]. Grains swell because of the combined effect of starch gelatinization, partial degradation of endosperm structure, solubilization of cellular wall, and partial solubilization of proteic matrix [27]. The peripheral and external endosperm also suffers small modifications because of the hardness of nixtamal grain [33]. Optimal cooking of masa is usually evaluated with the measurement of textural parameters such as plasticity, cohesivity, and chewiness [43]. Steeping promotes moisture diffusion inside the grain and produces nixtamal (cooked corn) that is homogeneously hydrated [44]. In the steeping period, swelling and solubility increase, nixtamalized corn grinding results in increased gelatinization and releasing of swollen granules [45]. In this period, the water absorption increases during the early hours because of pericarp removal (90%), allowing rapid diffusion of calcium in starch in the first 8 hours [46]. Cooking temperature, agitation, and average alkali concentration directly affect water uptake and calcium produced by interaction of alkaline solution with components in the corn kernel [47]. Several nutrients and antioxidants are lost, especially those present in yellow corn. It is recommended to select grains with soft endosperm to promote a more efficient action of calcium when interacts with starch pericarp [47, 48].

3.1.1.2. Washing

When nixtamal is washed, excess lime is removed and pH values are diminished, increasing loss of dry matter because of pericarp removal and color improvement observed in products. Some pentosane gums of corn are retained and are useful to maintain flexibility and smoothness in masa and tortillas [49]. Dry matter losses in the nixtamalization process, especially in alkaline cooking, steeping, and corn grain wash are between 5 and 14% (w/w) [50, 51]. Nejayote (cooking liquor) contains pericarp and soluble proteins, which are discharged every time when a batch is completed [52]. Soft endosperm, damaged grain, usage of high boiling temperatures, and excess of lime contribute to increase economic losses [43].

3.1.1.3. Grinding

Washed nixtamal is grounded in a stone mill to produce dough (masa) composed of different kinds of particles, including fragments of germ, pericarp, and endosperm, and other compounds such as entire starch granules, proteins, hydrated fibers, and fat, with 50–60% of humidity [45]. Amylopectin is solubilized as part of the mechanical fragmentation of corn kernel in stone milling [39]. The final masa particle size is affected by the grinding process. Masa particle size used for tortilla must be fine and coarse for snacks and corn chips, which is determined subjectively [53].

3.1.1.4. Mixing and masa formation

Mixing time and masa consistency are critical for a good machinability in tortilla disc-forming machine. During this step, the rheological and textural properties of masa are enhanced, acquiring the final quality that is required in the product [54]. This step includes calculation of dimensions and weight of final product. Soluble starch may increase because of the stone milling, affecting viscosity of starch and diminishing during masa processing [39]. During forming, the masa is rolled into a sheet, which is cut by a rotating cutter positioned underneath the rolls [2].

3.1.1.5. Tortilla baking

This step includes cooking and partial drying of masa, giving a light toast appearance to disc masa and developing final texture of the product. When masa is transformed into tortilla [39], crystallinity disappears or decreases to a higher extent than in the preceding processing steps [16].Temperature, steeping time, and lime concentration impact on textural quality of masa and tortilla [15]. Viscosity, cohesiveness, and adhesiveness of the tortillas are improved by alkaline cooking when corn masa is produced [55] and consequently appropriate texture results in good-quality tortilla. An appropriate texture gives adequate adhesivity when rolling machine is forming and cutting the tortilla. If corn is over-cooked, corn masa is sticky and adheres strongly to the rolls. Under-cooked corn produces a little cohesive masa and inadequate handling for tortilla disc formation [53].

3.2. Industrial nixtamalization process (INP) for corn flour production

The process used to produce NCF includes the following steps: reception, selection, cleaning, and storage, cooking and grinding of grain, dehydration, sifting, classification of product and packaging [56]. Dry masa is produced by drying and grinding lime-cooked and grounded masa. To obtain NCF, it is necessary to mill the nixtamal several times, using low-moisture content, and reducing contact time between starch and lime. These conditions in process affect deeply the structure of starch granule, avoiding the release of other components in corn kernel and changing the functionality of NCF [7].

Functionality of NCF is based on physicochemical characteristics, and it is increased after hydration process of corn masa is done [7]. A quality parameter used by masa manufacturers is water absorption content (WAC). Excessive heating is a major cause that damages starch granule, producing losses in its structure and integrity and producing flours with high WAC [36]. This is observed when certain quantity of water is retained by the hydration of NCF, which is related to a bigger economical profit. This behavior is related to the content of starch and protein levels in NCF particles [7].

The NCF presents a high particle size index (PSI), a quality parameter used to compare different types of flours [57]. The PSI is an indirect measurement of mechanical damage occurring in corn grain [58]. The flours are fractioned and reformulated for specific applications (table tortillas and fried products) [44]. The NCFs exhibit a more homogeneous particle size distribution (PSD) than does those for fried products [39]. The NCF milling is severe, and as a direct consequence, the content of damaged starch is increased when low gelatinization enthalpy values are registered. Textural quality in the NCF differs from TNP, and this behavior is governed by the following reasons:

a. A high value in particle size index (PSI) implies lower values of viscosity, hardness, and adhesivity because of a greater content of soluble solids [6, 59].

b. Water absorption in NCF is bigger than values observed in fresh masa, and more hydrogen bonds are present between starch chains and water molecules [39].

c. Retrogradation increases and shelf life shortens. More reorganization reactions occur because of the ratio and concentration of amylose and amylopectin [60].

d. Rapid drying of masa causes gelatinization and reorientation of starch polymers. A modification of the rheological characteristics of masa is observed when the baking of tortillas is performed, and lower rollability values and higher scores of hardness are registered [39].

When dry masa flour from the traditional process was compared against dry nixtamalized corn flour, gelatinization temperatures were similar; although enthalpy of gelatinization was lower in the second one as a consequence of a higher degree of gelatinization in corn starch [34, 36, 61].

3.3. Nixtamalization process by extrusion (NPE) to produce corn flour

Cooking extrusion is a continuous and a relative low-cost process used to modify the functional and digestible characteristics of cereal grains. All parts of corn grain are ground together, conditioned with lime and water (40–65% based on the weight of corn), and the mixture is heated with electric resistors or steam located at the outer side of the extruder [10]. Extrusion cooking is performed at high temperatures (90–120°C), low moisture, and short time, without the generation of contaminant effluents [62]. During the extrusion, changes in starch produce a higher degree of gelatinization as compared to TNP [1].

The main objective of producing extruded nixtamalized corn flours (ENCFs) is to modify the functional properties of starch. This objective is reached, modifying the extrusion parameters, such as temperature, feed moisture, and the speed of the screw [63–65]. In this way, starches with functional properties similar to those of chemically modified (hydroxypropylated and cross-linked) starches can be obtained without chemical modification. On the other hand, the functional properties of the extruded flours will depend on the extrusion conditions [66, 67]. This interaction is an important factor to consider in extruded products during single-screw extrusion [67–69].

Extrusion cooking modifies the starch crystallinity because the thermal treatment is more aggressive, and consequently, the crystalline structure of raw starch granule is partially or completely destroyed [70, 71]. The extruded nixtamalized corn flour can show a similar behavior to traditional process on gelatinization temperature (80°C) in accordance with adequate process variables observed [14, 17, 72]. There is a higher quantity of damaged starch in extruded corn starch as a consequence of the fragmentation caused by intense shear within the extruder [67, 68]. Loss of crystallinity is caused by mechanical disruption of the molecular bonds inside the starch structure, as a result of the extended dehydration [73].

Other process factors are important in extrusion cooking. Screw speed is considered a major factor, as also expansion index and density of extrudates [74]. A rise in pressure within the extruder is observed when the composition of raw matter changes [75]. Gómez-Aldapa et al. [76] observed that composition of raw matter used to make ENCF had a high impact in the production of tortillas.

The extrusion causes important nutritional changes in flours. A high humidity level and low temperature improve the nutritional characteristics of the treated products, while the more severe treatments (low humidity and high temperatures) make them worse. The treatment conditions can influence the generation of Maillard reactions, with consequent deterioration of the nutritional quality of the proteins. Nutritional composition of tortillas made via extrusion is better than those of the traditional process using integral grain. Gómez-Aldapa et al. [76] performed a nutritional comparison between two types of tortillas. They observed that tortillas made with ENCF had a higher content of dietary fiber and lysine than those prepared with nixtamal. They also concluded that pericarp and soluble solids discharged in cooking liquor diminished the nutritional value of tortillas made with fresh masa. The optimization process is used in extrusion to find the best combination of parameters to formulate a product. Milán-Carrillo et al. [77] established a set of parameters to produce tortillas with optimal conditions using high quantity of lysine and tryptophan maize. There are other formulations made with whole meal to increase nutritional value of flours [78, 79].

Extrusion changes the nutritional properties of nixtamalized products because of:

- **a.** Promotion of a more rapid transfer of water into molecules.
- **b.** The denaturation of proteins, reducing the lipid oxidation by inactivating the enzymes responsible for it. Lysine and protein content are increased.
- **c.** The complex formation and degradation of thermolabile vitamins and pigments.
- **d.** The increase of the soluble fiber content (expressed as damaged and resistant starch).
- **e.** The inactivation of enzymes (responsible for shortened shelf life in products in extrusion) and diminishing the microbial load [13].

The extruded flours produce changes in the rheological behavior of the starches. The involved modifications are similar to when the pastes are subjected to heating and cooling cycles [80]. However, extrusion causes starch changes more abrupt than traditional cooking methods, damaging a greater amount of starch granules and modifying the cold thickening power of the treated starches [81].

The extrusion has been used to produce corn flour and nixtamalized corn masa [82, 83]. However, this technology has not totally replaced the TNP [83]. González-Vera [72] produced tortillas using ENCF of high protein quality maize. The evaluation of the physicochemical changes of starch granule during the process was done. A harder product was produced, as a consequence of more severe damages in corn starch during the grinding and cooking stages.

The tortilla making process involves rupture of crystalline region of the starch granule and loss of molecular integrity producing a decrease in the intensity of the diffraction patterns due to the thermal treatment [16]. Arámbula-Villa et al. [73] produced tortillas using ENCF enriched with 3% (w/w) of pericarp. They observed that viscosity of ENCF was increased with the addition of maize pericarp. Using other concentrations, the viscosity decreased. The viscosity in ENCF is also increased during the production process with the addition of gums [3, 57] and enzymes [68]. The effect of alkaline cooking and lime concentration (from 0 to 1.0%) had direct impact on the viscosity of corn masa, when a Brabender amylograph was used to measure viscosity [6].

The milling of the extrudates and particle size distribution (PSD) of the flour are important parameters that define the functional properties of ENCF [57]. Changes in starch structure are affected by particle size [14]. Hasjim et al. [58] observed that the structure of starch in cereal grains is affected during milling, resulting in a great damage to starch granules. This phenomenon is dependent on milling conditions and type of equipment used. The mechanical damage in starch is different when a blade mill or a hammer mill is used in size reduction [17]. Damaged starch is related to small starch particles that hydrate easily when flour is produced. The higher damage in starch indicates the finest particle. A damaged starch has a higher water absorption capacity, which is good in industry, but too much starch damage leads to sticky and highly adhesive masa [84].

A problem of extruded flour is its low water absorption. The water absorption content (WAC) in ENCF is usually low, and corn masa loses water quickly, due to a high dehydration rate and retrogradation effects. Both factors result in harder tortillas [85]. Drying is important to eliminate water excess in the extrudates, and appropriate conditions are needed to avoid damage to starch and corn matter losses [50]. As an immediate consequence of drying process, starch is degraded and short chains are generated, retaining a higher number of water molecules [45]. Excessive heating affects deeply the structure and integrity of starch granule and will form a gelatinized paste with a higher content of dissolved solids [14]. WAC in masa can be improved using whole corn flour and a combination of ingredients [86].

Another characteristic related with the functionality of ENCF is amylose content [87]. Chinnaswamy and Hanna [65] characterized the macromolecular and functional properties of different corn starches. During extrusion cooking under various conditions, they concluded that starches rich in amylopectin (>50%) degrade faster than starches with a major amylose proportion. When the extrudates are produced at higher temperatures, water loss is more pronounced during cooling than those extruded at lower temperature [88]. Finally, Estrada-Girón et al. [37] evaluated the effect of the moisture content and temperature on the XRD, microstructural, pasting, thermal, and rheological properties of nixtamalized masa. They observed a high dependence of moisture content and temperature when physical and rheological properties were evaluated.

4. Texture and rheology of nixtamalized products

Texture is a sensory perception derived from the structure of food related with viscosity and elasticity [89]. Texture of masa is important because tortilla production is performed taking into account the masa consistency and is reflected in the formation of the sheet, favoring its cutting and shaping as round disks [38]. Texture is affected by endosperm texture, type of endosperm, drying process, storage life, and corn kernel variety, having a direct impact in the production of the different types of flours [90].

Textural properties of nixtamalized corn masa are dependent on the degree of gelatinization of the starch. In general, the corn masa obtained is defined as a mixture consisting of amylose and amylopectin mixed with partially gelatinized starch and intact granules, endosperm parts, and lipids [39]. An overcooked masa will be highly adhesive; meanwhile, an undercooked masa will have low cohesivity and bad machinability when tortillas are made [7]. Adhesiveness is a quality parameter in corn masa production evaluated instrumentally as the maximum tensile force during the adhesion process. It is defined as the cohesive rupture between two flat, circular metal plates, and the food sample [15].

The measurement of textural characteristics is evaluated using objective and subjective methods [91]. The objective methods or instrumental measurements use instruments that give numerical results in physical units and are independent of the operator. These tests can be fundamental, empiric, and imitative. The subjective methods evaluate the food quality using the sense of touch and, for this reason, are tough to standardize [91, 92]. The texture profile analysis (TPA) of masa, including parameters such as hardness, springiness, cohesiveness, and gumminess, is done using a texture analyzer [34]. Contreras-Jiménez et al. [1] evaluated ENCF and did not find differences when compared against the NCF. The hardness value range found in ENCF was between 1.01 and 3.15 N against nixtamal masa hardness value observed (1.79 N). ENCF is usually compared against the masa prepared by TNP, using the right formulation to get similar results in texture. A better texture value in nixtamal is attributed to appropriate swelling of the starch granules, the hydrolysis of the pericarp that release gums from the nixtamalized pericarp (hemicellulose) and the presence of saponified lipids (used in a natural way as enhancers of texture) in the germ [93].

There are several rheological methods to measure texture in corn masa and slurries such as back extrusion [89, 94], squeezing flow viscometry, creep test [95], and dynamic rheology [96]. Dynamic rheology has measured efficiently the viscoelastic properties of materials, especially starch [57]. Sahai et al. [34] found a strong relationship between textural attributes of masa obtained from TPA, the particle size, and the composition. Arámbula-Villa et al. [17] evaluated some textural characteristics of tortillas prepared with four types of gums added before and after extrusion and established ideal textural parameters in the tortilla. Corradini and Peleg [97] applied squeezing flow viscometry to measure the rheology of several semiliquid foods. This method has been useful to characterize corn masa, evaluating rheological properties and monitoring their ability to recover original consistency after a shearing is applied [15].

The measurement of viscoelastic characteristics of masa has been applied successfully to describe the behavior of corn starch using a range that implies a small deformation in the test material. There are two concepts useful to understand viscoelasticity. First, the storage modulus (G') is referred to elastic modulus. Lost modulus (G″) is related to the viscous modulus. Vázquez-Carrillo et al. [98] found a great dependence of G' and G″ on frequency. They observed that G' was always higher than G″, which is a characteristic behavior of starch gels. Méndez-Montealvo et al. [42] observed similar behavior when they measured G' and G″ in nixtamalized maize starch at two stages (90 and 25°C) of the starch gelation. Mondragón et al. [5]

used small amplitude oscillatory rheometry to study the influence of lime and amylose-lipid complexes on the viscoelastic behavior of nixtamalized maize starch gels and observed that G' and G″ showed to be lime dependent. Platt-Lucero et al. [57] studied the viscoelastic behavior of masa obtained from rehydrated extruded corn flour and observed that G'>G″ at any frequency and for any treatment. Quintanar-Guzmán et al. [38] observed a similar behavior when the viscoelastic properties of corn were evaluated under different nixtamalization conditions.

The measurement of physicochemical properties, especially the moisture content is an important issue to consider. The raw tortilla is highly susceptible to moisture loss. The moisture content in fresh masa and tortilla represents around 40–50% of total weight. An excessive loss of water in final product is related to economic losses, which is unacceptable for producers. High moisture content in tortilla is related to a high microbiological charge causing a brief shelf life and an increase in starch retrogradation [33, 99]. As an immediate consequence, a problem very common in tortillas prepared with masa obtained from ENCF is the gradual increase in hardness [7, 36].

Other physical characteristics evaluated in tortilla are weight, diameter, and thickness. These parameters in tortillas produced via extrusion [57] were similar when compared to those of the traditional nixtamalization process [15]. Tortilla texture includes the measurement of hardness and rollability during the first 48 h. ENCF used to make masa, and thereafter, tortillas can be mixed with gums to increase its WAC and tortilla yield [17]. The effect of extrusion conditions continues to be studied. Chaidez-Laguna et al. [100] emphasized the role of protein content in corn tortilla made with ENCF and concluded that mixing time is critical to enhance the consistency of flour and the texture in the tortilla. Reyes-Moreno et al. [101] improved the end quality and nutritional value of product, since it is very promising to use whole corn grain in the nixtamalization via extrusion.

5. Conclusions

Discussion has been focused in a comparison between the characteristics of several nixtamalization processes. Advantages and disadvantages of nixtamalization and extrusion were discussed. Functional properties are different when starch is processed by diverse conditions. Best textural quality is achieved using traditional alkaline cooking, but it has a wide variation in control process; it produces contaminant effluents, takes long time to be performed, and the production at commercial levels is small. Usage of NCF as an alternative nixtamalization process has proved to be an economic and reliable option to make tortillas. Processing of NCF yields a product with a regular textural quality that stale and mold faster. NCF technology also produces contamination as the traditional way does. Extrusion is a continuous process involving low-feed moisture level, an adequate cooking temperature, and a correct lime addition to produce whole corn extrudates, flour, and then tortillas. Extrusion can be applied correctly handling the process variables to offer an alternative to make corn flours and diminishing a more severe damage in corn starch. Extrusion behaves similarly to traditional way and has approximately the same product quality. Extrusion is performed without pollution, is efficient, and increases nutritional value of corn when adequate raw matter is chosen. There are researchers that support the use of extrusion as an alternative to produce tortilla, and it has been a necessity detected in food industry.

Acknowledgements

Thanks to CONACyT for doctoral scholarship support.

Nomenclature

NCF	nixtamalized corn flour
MW	molecular weight
TIA	Tortilla Industry Association
TNP	traditional nixtamalization process
INP	industrial nixtamalization process
NPE	nixtamalization process by extrusion
"d"	interplanar spacing
ENCF	extruded nixtamalized corn flour
WAC	water absorption content
PSI	particle size index
PSD	particle size distribution
TPA	texture profile analysis
G'	storage modulus
G"	lost modulus

Author details

Carlos Martín Enríquez Castro[1], Patricia Isabel Torres-Chávez[2], Benjamín Ramírez-Wong[2]*, Ana Irene Ledezma-Osuna[2], Armando Quintero-Ramos[3], Jaime López-Cervantes[4] and María Irene Silvas-García[2]

*Address all correspondence to: bramirez@guaymas.uson.mx

1 Food Science and Technology Program, University of Sonora, Hermosillo, Sonora, Mexico

2 University of Sonora, Hermosillo, Sonora, Mexico

3 Autonomous University of Chihuahua, Chihuahua, Mexico

4 Technological Institute of Sonora, Obregón, Sonora, Mexico

References

[1] Contreras-Jiménez B, Morales-Sánchez E, Reyes-Vega ML, Gaytán-Martínez M. Functional properties of extruded corn Flour obtained at low temperature. CytA-Journal of Food. 2014;**12**:263-270

[2] Serna-Saldivar, SO. Química, almacenamiento e industrialización de los cereales. SA México, DF: AGT Editor; 1996. pp. 408-420

[3] Serna-Saldivar, SO, Gómez, MH, Rooney, LW. Technology, chemistry, and nutritional value of alkaline-cooked corn products. Advances in Cereal Science and Technology. 1990;**10**:243-306

[4] Gaytán-Martínez M, Figueroa-Cárdenas JD, Morales-Sánchez E, Vázquez-Landaverde PA, Martínez-Flores HE. Physicochemical properties of masa and corn tortilla made by ohmic heating. African Journal of Biotechnology. 2011;**10**:16028-16036

[5] Mondragón, M, Bello-Pérez, LA, Agama-Acevedo, E, Betancourt-Ancona, D, Peña JL. Effect of cooking time, steeping and lime concentration on starch gelatinization of corn during nixtamalization. Starch-Stärke. 2004;**56**:248-253

[6] Bryant CM, and Hamaker BR. Effect of lime on gelatinization of corn flour and starch. Cereal Chemistry. 1997;**74**:171-175

[7] Bello-Pérez, LA, Osorio-Díaz, P, Agama-Acevedo, E, Núñez-Santiago, C, Paredes-López, O. Chemical, physicochemical and rheological properties of masas and nixtamalized corn flour. Agrociencia. 2002;**36**:319-328

[8] Gutiérrez-Uribe JA, Rojas-García C, García-Lara S, Serna-Saldivar SO. Phytochemical analysis of wastewater (nejayote) obtained after lime-cooking of different types of maize kernels processed into masa for tortillas. Journal of Cereal Science. 2010;**52**:410-416

[9] Bressani R, Paz y Paz R, Scrimshaw NS. Corn nutrient losses, chemical changes in corn during preparation of tortillas. Agricultural and Food Chemistry. 1958;**6**:770-774

[10] Harper JM Extrusion of foods. Volume I. CRC Press Inc. 2000 Corporate Blvd, N.W., Boca Raton, FL USA; 1981; 212 p.

[11] Zhu S, Riaz MN, Lusas EW. Effect of different extrusion temperatures and moisture content on lipoxygenase inactivation and protein solubility in soybeans. Journal of Agriculture and Food Chemistry. 1996;**44**:3315-3318

[12] Guy, RCE. Extrusion cooking technologies and applications. 1st ed. CRC Press. N.W., Boca Raton, FL USA; 2001; 200 p.

[13] Rendón-Villalobos R, Bello-Pérez LA, Osorio-Díaz P, Tovar J, Paredes-López O. Effect of storage time on in vitro digestibility and resistant starch content of nixtamal, masa, and tortilla. Cereal Chemistry. 2002;**79**:340-344

[14] Lai LS, Kokini JL. Physicochemical changes and rheological properties of starch during extrusion. (A review). Biotechnology Progress. 1991;**7**:251-266

[15] Ramírez-Wong B, Sweat VE, Torres-Chávez PI, Rooney LW. Development of two instrumental methods for corn masa texture evaluation. Cereal Chemistry. 1993;**70**:286-290

[16] Campus-Baypoli ON, Rosas-Burgos EC, Torres-Chávez PI, Ramírez-Wong B, Serna-Saldivar SO. Physicochemical changes of starch during maize tortilla production. Starch-Stärke. 1999;**51**:173-176

[17] Arámbula-Villa G, Mauricio SRA, Figueroa-Cárdenas JD, González-Hernández J, Ordorica FCA. Corn masa and tortillas from extruded instant corn flour containing hydrocolloids and lime. Journal of Food Science. 1999;**64**:120-124

[18] Ramirez-Wong B, Sweat VE, Torres-Chávez, PI, Rooney LW. Cooking time, grinding, and moisture content effect on fresh corn masa texture. Cereal Chemistry. 1994;**71**:337-343

[19] French D. Organization of starch granules. Starch: Chemistry and Technology. 1984;**2**: 183-247

[20] BeMiller J N, Whistler R L. Starch: Chemistry and Technology. 3rd ed. Academic Press. CA, USA; 2009.

[21] Thompson DB. Strategies for the manufacture of resistant starch. Trends in Food Science and Technology. 2000;**11**:245-253

[22] Kossman J, Lloyd J. Understanding and influencing starch biochemistry. Critical Reviews in Plant Science. 2000;**19**:171-226

[23] Bello-Pérez LA, Rodríguez-Ambriz SL, Sánchez-Rivera MM, Agama-Acevedo E. Starch macromolecule structure. In: Bertolini AC. Starches: Characterization, Properties, and Applications. Boca Raton: Taylor and Francis. 2010. pp. 33-59

[24] Mua JP, Jackson DS. Relationships between functional attributes and molecular structures of amylose and amylopectin fractions from corn starch. Journal of Agricultural and Food Chemistry. 1997;**45**:3848-3854

[25] Donald AM. Understanding starch structure and functionality. In: Eliasson AC. (Ed). Starch in food: structure, function and applications. CRC Press. 2004. Corporate Blvd, N.W., Boca Raton, FL USA; 2000; pp. 158-179.

[26] Biliaderis CG, Zawistowski J. Viscoelastic behavior of aging starch gels: Effects of concentration, temperature, and starch hydrolysates on network properties. Cereal Chemistry. 1990;**67**:240-246

[27] Buléon A, Colonna P, Planchot V, Ball S. Starch granules: Structure and biosynthesis. International Journal of Biology of Macromolecules. 1998;**23**:85-112

[28] Hill RD, Dronzek BL. Scanning electron microscopy studies of wheat, potato and corn starch during gelatinization. Starch-Stärke, 1973;**25**:367-372

[29] Gidley MJ. Starch structure/function relationships: Achievements and challenges. In: Barsby TL, Donald AM, Frazier PJ. Starch: Advances in Structure and Function. The

Proceedings of Starch 2000: Structure and function held on 27-29 March 2000 at Churchill College, Cambridge. pp. 1-4

[30] Tester RF, Debon SJJ. Annealing of starch: A review. International Journal of Biological, Macromolecules. 2000;**27**:1-12

[31] Figueroa-Cárdenas JD, Véles-Medina JJ, Hernández-Landaverde MA, Aragón-Cuevas F, Gaytán-Martínez M, Chávez-Martínez E, Palacios N, Willcox M. Effect of annealing from traditional nixtamalisation process on the microstructural, thermal, and rheological properties of starch and quality of pozole. Journal of Cereal Science. 2013;**58**:457-464

[32] Tortilla Industry Association [Internet]. 2016. Available from: http://tortilla-info.com [Accessed: 20-09-2016]

[33] Gómez MH, McDonough CM, Rooney LW, Waniska RD. Changes in corn and sorghum during nixtamalization and tortilla baking. Journal of Food Science. 1989;**54**:330-336

[34] Sahai D, Mua JP, Surjewan I, Buendía MO, Rowe M, Jackson DS. Alkaline processing (nixtamalisation) of white Mexican corn hybrids for tortilla production: Significance of corn physicochemical characteristics and process conditions 1. Cereal Chemistry. 2001;**78**:116-120

[35] Mariscal-Moreno RM, Figueroa-Cárdenas JD, Santiago-Ramos D, Arámbula-Villa GA, Jiménez-Sandoval S, Rayas-Duarte P, Véles-Medina JJ, Martínez-Flores HE. The effect of different nixtamalization processes on some physicochemical properties, nutritional composition and glycemic index. Journal of Cereal Science. 2015;**65**:140-146

[36] Bello-Pérez LA, Osorio-Díaz P, Agama-Acevedo E, Solorza-Feria J, Toro-Vázquez J, Paredes-López O. Chemical and physicochemical properties of dried wet masa and dry masa flour. Journal of Science Food Agriculture. 2003;**83**:408-412

[37] Estrada-Girón Y, Aguilar J, Morales-del Rio JA, Valencia-Botín AJ, Guerrero-Beltrán JA, Martínez-Preciado AH, Fernández VVA. Effect of moisture content and temperature on the rheological, microstructural and thermal properties of masa (dough) from a hybrid corn (*Zea Mays sp.*) variety. Revista Mexicana de Ingeniería Química. 2014;**13**:429-446

[38] Quintanar-Guzmán A, Flores MEJ, Escobedo RM, Guerrero LC, Feria JS. Changes on the structure, consistency, physicochemical and viscoelastic properties of corn (Zea mays sp.) under different nixtamalisation conditions. Carbohydrate Polymers. 2009;**78**:908-916

[39] Gómez MH, Lee JK, McDonough CM, Waniska RD, Rooney LW. Corn starch changes during tortilla and tortilla chip processing. Cereal Chemistry. 1992;**69**:275-279

[40] Ratnayake WS, Wassinger AB, Jackson DS. Extraction and characterization of starch from alkaline cooked corn masa. Cereal Chemistry. 2007;**84**:415-422

[41] Méndez-Montealvo G, Bello-Pérez LA, Solorza-Feria J, Velázquez-del Valle M, Montiel N, Paredes-López O. Composición química y caracterización calorimétrica de híbridos y variedades de maíz cultivadas en México. Agrociencia. 2005;**39**:267-274

[42] Méndez-Montealvo G, Sánchez-Rivera MM, Paredes-López O, Bello-Pérez LA. Thermal and rheological properties of nixtamalised maize starch. International Journal of Biological Macromolecules. 2006;**40**:59-63

[43] Pflugfelder RL, Rooney LW, Waniska RD. Fractionation and composition of commercial corn masa. Cereal Chemistry. 1988;**65**:262-266

[44] Almeida-Domínguez HD, Domínguez-Cepeda M, Rooney LW. Properties of commercial nixtamalized corn flours. Cereal Foods World. 1996;**41**:624-630

[45] Almeida-Domínguez HD, Suhendro EL, Rooney LW. Corn alkaline cooking properties related to grain characteristics and viscosity (RVA). Journal of Food Science. 1997;**62**: 516-519

[46] Gutiérrez-Uribe JA, Rojas-García C, García-Lara S, Serna-Saldivar SO. Effects of lime-cooking on carotenoids present in masa and tortillas produced from different types of maize. Cereal Chemistry. 2014;**91**:508-512

[47] Ruiz-Gutiérrez MG, Quintero-Ramos A, Meléndez-Pizarro CO, Talamás-Abbud R, Barnard J, Márquez-Meléndez R, Lardizábal-Gutiérrez D. Nixtamalisation in two steps with different calcium salts and the relationship with chemical, texture and thermal properties in masa and tortilla. Journal of Food Process Engineering. 2012;**35**:772-783

[48] Gutiérrez-Cortez E, Rojas A, Cornejo-Villegas MA, Zepeda-Benitez Y, Rodríguez-García ME. Microstructural changes in the maize kernel pericarp during cooking stage in nixtamalisation process. Cereal Science. 2010;**51**:81-88

[49] Carvajal-Millan E, Rascón-Chu A, Márquez-Escalante JA, Micard V, Ponce de León N, Gardea A. Maize bran gum: Extraction, characterization and functional properties. Carbohydrate Polymers. 2007;**69**:280-285

[50] Pflugfelder RL, Rooney LW, Waniska RD. Dry matter losses in commercial corn masa production. Cereal Chemistry. 1988;**65**:127-132

[51] Rooney LW, Suhendro EL. Perspectives on nixtamalisation (alkaline cooking) of maize for tortillas and snacks. Cereal Foods World. 1999;**44**:466-470

[52] Rosentrater KA. A review of corn masa processing residues: Generation, properties, and potential utilization. Waste Manage. 2006;**26**:284-292

[53] Ramirez-Wong B, Sweat VE, Torres-Chávez PI, Rooney LW. Cooking time, grinding, and moisture content effect on fresh corn masa texture. Cereal Chemistry. 1994;**71**:337-343

[54] Bello-Perez LA, Paredes-López, O. Starches of some food crops, changes during processing and their nutraceutical potential. Food Engineering Reviews. 2009;**1**:50-65

[55] Martín-Martínez ES, Jaime-Fonseca MR, Martínez-Bustos F, Martínez-Montes JL. Selective nixtamalisation of fractions of maize grain (*Zea mays L.*) and their use in the preparation of instant tortilla flours analyzed using response surface methodology. Cereal Chemistry. 2003;**80**:13-19

[56] Rooney LW, Suhendro EL. Perspectives on nixtamalisation (alkaline cooking) of maize for tortillas and snacks. Cereal Foods World. 1999;**44**:466-470

[57] Platt-Lucero LC, Ramírez-Wong B, Torres-Chávez PI, López-Cervantes J, Sánchez-Machado, DI, Reyes-Moreno C, Morales-Rosas I. Improving textural characteristics of tortillas by adding gums during extrusion to obtain nixtamalised corn flour. Journal of Texture Studies. 2010;**41**:736-755

[58] Hasjim J, Li E, Dhital S. Milling of rice grains: The roles of starch structures in the solubility and swelling properties of rice flour. Starch-Stärke. 2012;**64**:631-645

[59] Sahai D, Buendía MO, Jackson DS. Analytical techniques for understanding nixtamalized corn flour: Particle size and functionality relationships in a masa flour sample. Cereal Chemistry. 2001;**78**:14-18

[60] Paredes-López O, Saharópulos-Paredes ME. Scanning electron microscopy studies in limed corn. Journal of Food Science Technology. 1983;**17**:687-693

[61] Wen LF, Rodis P, Wasserman BP. Starch fragmentation and protein insolubilization during twin-screw extrusion of corn meal. Cereal Chemistry. 1990;**67**:268-275

[62] Harper JM. Extrusion of foods. In: IFT Symposium Series: Biotechnology and Food Process Engineering; 1990. pp. 295-308

[63] Camire ME, Camire A, Krumhar K. Chemical and nutritional changes in foods during extrusion. Critical Reviews in Food Science and Nutrition. 1990;**29**:35-57

[64] Charbonniere R, Duprat F, Guibolt A. Changes in various starches by extrusion cooking 2: Physical structures of extruded products. Cereal Science Today. 1973;**l8**:286

[65] Chinnaswamy R, Hanna MA. Macromolecular and functional properties of native and extrusion-cooked corn starch. Cereal Chemical 1990;**67**:490-499

[66] Kljak K, Šárka E, Dostálek P, Smrčková P, Grbeša D. Influence of physicochemical properties of Croatian maize hybrids on quality of extrusion cooking. LWT-Food Science and Technology. 2015;**60**:472-477

[67] Curic D, Novotni D, Bauman I, Kricka T, Dugum J. Optimization of extrusion cooking of cornmeal as raw material for bakery products. Journal of Food Process Engineering. 2009;**32**:294-317

[68] Platt-Lucero LC, Ramírez-Wong B, Torres-Chávez PI, López-Cervantes J, Sánchez-Machado DI, Carvajal-Millán E, Morales-Rosas I. Effect of xylanase on extruded nixtamalised corn flour and tortilla: Physicochemical and rheological characteristics. Journal of Food Process Engineering. 2013;**36**:179-186

[69] Bhattacharya M, Hanna MA. Kinetics of starch gelatinization during extrusion cooking. Journal of Food Science 1987;**52**:764-766

[70] Lawton BT, Henderson GA, Derlatka EJ. The effects of extruder variables on the gelatinization of corn starch. Canadian Journal of Chemical Engineering. 1972;**250**:168-172

[71] Bazúa CD, Guerra R, Sterner H. Extruded corn flour as an alternative to lime-heated corn flour for tortilla preparation. Journal of Food Science. 1979;**44**:940-941

[72] González-Vera I. Evaluación de cambios fisicoquímicos que sufre el almidón en el proceso de elaboración de tortillas, utilizando harinas nixtamalizadas por extrusión de maíz de alta calidad proteica (Thesis). Hermosillo, Sonora, México: University of Sonora; 2006

[73] Arámbula-Villa G, González-Hernández J, Moreno ME, Ordorica Falomir CA. Characteristics of tortillas prepared from dry extruded masa flour added with maize pericarp. Journal of Food Science. 2002;**67**:1444-1448

[74] Gómez-Aldapa C, Martínez-Bustos F, Figueroa-Cárdenas JD, Ordorica FCA. A comparison of the quality of whole corn tortillas made from instant corn flours by traditional or extrusion processing. International Journal of Food Science Technology. 1999;**34**:391-399

[75] Kokini JL, Baumann GC, Bresslauer K, Chedid LL, Herh P, Lai LS, Medeka H. A kinetic model for starch gelatinization and effect of starch/protein interactions on rheological properties of 98% amylopectin and amylose rich starches. Engineering and Foods: Advanced Process. Spiess, Schubert, Eds; Elsevier / Applied Science Publishers, NY USA 1990; Vol. 1, pp. 109-121.

[76] Gómez-Aldapa C, Martínez-Bustos F, Figueroa-Cárdenas JD, Ordorica-Falomir CA, González Hernández J. Chemical and nutritional changes during preparation of whole corn tortillas prepared with instant flour obtained by extrusion process. Archivo Latinoamericano de Nutrición. 1996;**46**:315-319

[77] Milán-Carrillo J, Gutiérrez-Dorado R, Perales-Sánchez JXK, Cuevas-Rodríguez EO, Ramírez-Wong B, Reyes-Moreno C. The optimization of the extrusion process when using maize flour with a modified amino acid profile for making tortillas. International Journal of Food Science and Technology. 2006;**41**:727-736

[78] Balandran-Quintana RR, Barbosa-Canovas GV, Zazueta-Morales JJ, Anzaldúa-Morales A, Quintero-Ramos A. Functional and nutritional properties of extruded whole pinto bean meal (*Phaseolus vulgaris* L.). Journal of Food Science. 1998;**63**:113-116

[79] Pérez AA, Drago SR, Carrara CR, De Greef, DM, Torres RL, González RJ. Extrusion cooking of a maize/soybean mixture: Factors affecting expanded product characteristics and flour dispersion viscosity. Journal of Food Engineering. 2008;**87**:333-340

[80] Hagenimana A, Ding X, Fang T. Evaluation of rice flour modified by extrusion cooking. Journal of Cereal Science. 2006;**43**:38-46

[81] Wolf B. Polysaccharide functionality through extrusion processing. Current Opinion in Colloid and Interface Science. 2010;**15**:50-54

[82] Martinez-Bustos F, Sanchez-Sinencio F, Gonzalez-Hernandez J, Martinez JD, Ruiz-Torres M. U.S. Patent No. 5532013. Washington, DC: U.S. Patent and Trademark Office; 1996

[83] Gutiérrez-Dorado R, Ayala-Rodríguez AE, Milán-Carrillo J, López-Cervantes J, Garzón-Tiznado JA, López-Valenzuela JA, Reyes-Moreno C. Technological and nutritional properties of flours and tortillas from nixtamalized and extruded quality protein maize (*Zea mays L.*). Cereal Chemistry. 2008;**85**:808-816

[84] Ghodke SK, Ananthanarayan L, Rodríguez L. Use of response surface methodology to investigate the effects of milling conditions on damaged starch, dough stickiness and chapatti quality. Food Chemistry. 2009;**112**:1010-1015

[85] Arámbula-Villa G, González-Hernández J, Ordorica-Falomir CA. Physicochemical, structural and textural properties of tortillas from extruded instant corn flour supplemented with various types of corn lipids. Journal of Cereal Science. 2001;**33**:245-252

[86] Al-Okbi SY, Hussein A, Hamed IM, Mohamed DA, Helal AM. Chemical, rheological, sensorial and functional properties of gelatinized corn-rice bran flour composite corn flakes and tortilla chips. Journal of Food Processing and Preservation. 2014;**38**:83-89

[87] Colonna P, Tayeb J, Mercier C. Extrusion cooking of starch and starchy products. In: Mercier C, Linko P, Harper LM editors. Extrusion Cooking. USA: AACC; 1989. pp. 247-319

[88] Arámbula-Villa G, Figueroa-Cárdenas JD, Martinez-Bustos F, Ordorica-Falomir CA, Gonzalez-Hernandez J. Milling and processing parameters for corn tortillas from extruded instant dry masa flour. Journal of Food Science. 1998;**63**:338-341

[89] Ignacio RM, Lannes SC. Rheological characterization and texture of commercial mayonnaise using back extrusion. African Journal of Agriculture Research. 2013;**8**:4262-4268

[90] Limanond B, Castell-Perez E, Moreira RG. Effect of time and storage conditions on the rheological properties of masa for corn tortillas. LWT-Food Science Technology. 1999;**32**:344-348

[91] Steffe J F. Rheological methods in food process engineering. 2nd ed. Freeman Press. 2807 Still Valley Dr. East Lansing MI USA 1996; pp. 255-258, 294-310.

[92] Karim AA, Norziah MH, Seow CC. Methods for the study of starch retrogradation. Food Chemistry. 2000;**71**:9-36

[93] Flores-Farías R, Martínez-Bustos F, Salinas-Moreno Y, Chang YK, Hernández JG, Ríos E. Physicochemical and rheological characteristics of commercial nixtamalised Mexican maize flours for tortillas. Journal of Science Food Agriculture. 2000;**80**:657-664

[94] Ramaswamy HS, Singh A, Sharma M. Back extrusion rheology for evaluating the transitional effects of high pressure processing of egg components. Journal of Texture Studies. 2015;**46**:34-45

[95] Ditudompo S, Takhar PS, Ganjyal GM, Hanna MA. The effect of temperature and moisture on the mechanical properties of extruded corn starch. Journal of Texture Studies. 2013;**44**(3):225-237

[96] Engmann J, Servais C, Burbidge AS. Squeeze flow theory and applications to rheometry: A review. Journal of Non-Newtonian Fluid Mechanics. 2005;**132**:1-27

[97] Corradini MG, Peleg M. Consistency of dispersed food systems and its evaluation by squeezing flow viscometry. Journal of Texture Studies. 2005;**36**:605-629

[98] Vázquez-Carrillo MG, Santiago-Ramos D, Gaytán-Martínez M, Morales-Sánchez E, Guerrero-Herrera JM. High oil content maize: Physical, thermal and rheological properties of grain, masa, and tortillas. LWT-Food Science Technology. 2015;**60**:156-161

[99] Ramírez-Wong B, Walker CE, Ledesma-Osuna AI, Torres-Chávez PI, Medina-Rodríguez CL, López-Ahumada GA, Flores RA. Effect of flour extraction rate on white and red winter wheat flour compositions and tortilla texture. Cereal Chemistry. 2007;**84**:207-213

[100] Chaidez-Laguna LD, Torres-Chavez PI, Ramírez-Wong B, Marquez-Ríos E, Islas-Rubio AR, Carvajal-Millan E. Corn proteins solubility changes during extrusion and traditional nixtamalisation for tortilla processing: A study using size exclusion chromatography. Journal of Cereal Science. 2016;**69**:351-357

[101] Reyes-Moreno C, Ayala-Rodríguez AE, Milán-Carrillo J, Mora-Rochín S. Production of nixtamalized flour and tortillas from amaranthine transgenic maize lime-cooked in a thermoplastic extruder. Journal of Cereal Science. 2013;**58**:465-471

3

Design of Polymer Extrusion Dies Using Finite Element Analysis

G.N. Kouzilos, G.V. Seretis, C.G. Provatidis and D.E. Manolakos

Abstract

A computational fluid dynamics (CFD) model has been developed to compute the pressure, temperature, velocity, viscosity and viscous dissipation in the high-density polyethylene (HDPE) extrusion process. The numerical approach agrees fairly well with the experimental data recorded during the extrusion process of the material. The extrusion spider die was designed to produce high-density polyethylene pipes of 32 mm inner nominal diameter and 2.4 mm thickness. In order to investigate if the spider legs are able to perform under the pressure occurred using the maximum flow rate provided by the single screw extruder of this study, a stress analysis was conducted on a single spider leg. This fluid-structure interaction (FSI) problem was solved using the COMSOL Multiphysics software. Finally, the results obtained from the FE analysis were applied in the design and fabrication of the spider die, selecting IMPAX (tool steel) as fabrication material.

Keywords: finite element analysis, pressure flow, HDPE, extrusion die, spider, arbitrary Lagrangian-Eulerian (ALE)

1. Introduction

The production of extruded polyethylene film, rods, tubes and pipes is a common industrial process that has been the subject of major investigations over many years [1–3]. The designing of extrusion dies for the production of such geometries requires a deep knowledge. It is usually based on experimental trial-and-error approaches, involving, therefore, the use of huge amounts of time and material resources [4–6]. According to manufacturers, 10–15 iterations are required to optimize complex profile geometries [1]. The extrusion die is one of the most important parts in extrusion processing. The extrusion die design process can become too difficult to execute, or its cost can increase up to prohibitive levels, when complex geometry thermoplastic profiles are concerned.

Optimizing process parameter problems is routinely performed in the manufacturing industry, particularly in setting final optimal process parameters. Final optimal process parameter setting is recognized as one of the most important steps in plastics extrusion for improving the quality of extruded products.

Yilmaz et al. [7] optimized the geometric parameters of a profile extrusion die, using several objective function definitions by Simulated Annealing-Kriging Meta-Algorithm. Objective functions are defined based on the uniformity of velocity distribution at the die exit. For this, computational fluid dynamics (CFD) simulations are performed for N = 70 die geometries. Appropriate geometric parameters (t and L) of the die are variables for the optimization problem.

The optimization of an extrusion die designed for the production of a wood-plastic composite (WPC) decking profile is investigated by Gonçalves et al. [8]. The optimization was performed with the help of numerical tools, more precisely, by solving the continuity and momentum conservation equations that govern such a flow, and aiming to balance properly the flow distribution at the extrusion die flow channel outlet.

A nonlinear optimization technique was applied by Mamalis et al. [9] to the numerical model to pinpoint the processing conditions, namely inlet pressure, inlet temperature of the melt, temperature of the die walls and temperature of the spider legs.

The work described, hereinafter, is aiming to the development of an optimum design for a pipe die with spider used for the extrusion of high-density polyethylene (HDPE) tubes. For this purpose, a computational fluid dynamics (CFD)-based model using the generalized Newtonian approach was employed to investigate pressure drop, flow and temperature uniformity in the die.

2. Extrusion die design zones

In order to determine the die pressure, that is, the pressure developed in the inner surfaces of the die, the analytical approach, which is presented below, was used. The extrusion die was considered to consist of five different zones. In each zone, a different stage of the extrusion process was taking place. In zone 1, the fluid enters the die (input or inlet). In zone 2, the fluid diverts from the extrusion axis. In this stage of the extrusion process, the distribution of the fluid begins on the top of the mandrel cone and, subsequently, the fluid is driven to zone 3 through a ring-shaped leak. In zone 3, the fluid is leaking in the spider legs, which are fitting in the male end of the die. A relaxation zone (zone 4) follows zone 3. The last stage is zone 5,

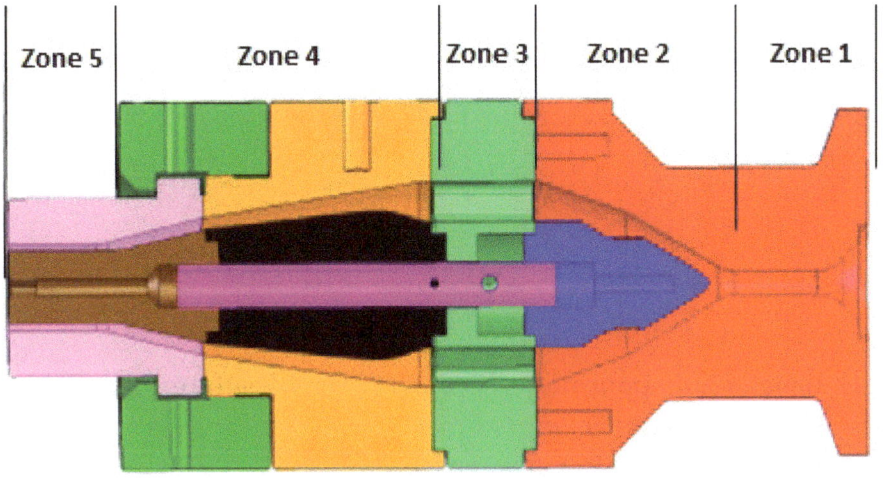

Figure 1. Zones of the extrusion die.

where the pipe is being formed at its expected morphological characteristics. The extrusion die zones are presented schematically in **Figure 1**.

3. Mathematical model

In order to determine the pressure drop in the extrusion die, the power-law exponential model was used, according to which the volumetric flow rate $\dot{V}$ of a non-Newtonian fluid is described by Eqs. (1) and (2) [1, 2].

$$\dot{V} = K' \cdot \varphi \cdot \Delta P^m \tag{1}$$

where φ is the fluidity, m is the flow exponent, ΔP is the pressure drop and K' is a die shape constant.

For the present work, a single-screw Johnson Plastics extruder was used to drive the flowing high-density polyethylene into the spider die. For this kind of extruder, the volumetric flow rate, as it has been described previously [2], is

$$\dot{V} = \alpha \cdot \mathrm{N} - \frac{\beta}{\mu}\frac{\Delta P}{L} \tag{2}$$

where $\alpha = 0.5 \cdot \pi^2 \cdot D^2 \cdot H \cdot \sin\phi \cdot \cos\phi$, N is the screw speed, $\beta = \frac{\pi}{12} D \cdot H^3 \cdot \sin^2\phi$, μ is the melt viscosity, L is the axial length of the screw, D is the inner barrel diameter, H is the depth of the channel and φ is the helix angle of flight [10].

In the five different zones of the extrusion die, three shapes of the cross section can be found: tube, ring shape and square shape. For these cross sections, the pressure drop is described by Eqs. (3)–(5), respectively [1, 10].

$$\Delta P_{tube} = L\left[\frac{2^{m}(m+3)\,\dot{V}}{\varphi\cdot\pi\cdot R^{m+3}}\right]^{1/m} \tag{3}$$

$$\Delta P_{ring} = L\left[\frac{2^{m+1}(m+2)\,\dot{V}}{\phi\cdot\pi\cdot D\cdot H^{m+2}}\right]^{1/m} \tag{4}$$

$$\Delta P_{square} = L\left[\frac{2^{m+1}(m+2)\,\dot{V}}{\phi\cdot B\cdot H^{m+2}}\right]^{1/m} \tag{5}$$

These equations have broad applications because the flow path in a small segment of many extrusions dies and adaptors can be approximated by a circular tube or a slit for the purpose of calculating pressure drop and flow rate. For a zero value of ΔP, the volumetric flow rate is maximized. Thus, for screw speed equal to 100 rpm, the maximum volumetric flow rate can be calculated equal to 7.9×10^{-6} m^3/s.

For this flow rate, the total drop of the pressure in the die ΔP_T, including all five different zones, is:

$$\Delta P_T = \Delta P_{Zone1} + \Delta P_{Zone2} + \Delta P_{Zone3} + \Delta\ P_{Zone4} + \Delta P_{Zone5} = 17.9\ bar + 9.55\ bar + 1.58\ bar + 27.34\ bar + 35.8 = 92.17\ bar \tag{6}$$

If ΔP_T is the total drop of the die pressure and ρ and C_p are the density and the specific heat, respectively, then the average temperature increase at the die output (outlet), which is based on the assumption that adiabatic conditions occur throughout the whole process, can be expressed by Eq. (7) [2].

$$\Delta T_{analytical} = \frac{\Delta P}{\rho\cdot C_p} = 4.72\ \text{K} \tag{7}$$

4. Design of the extrusion die

The extrusion die was designed to produce high-density polyethylene pipes of 32 mm inner nominal diameter and 2.4 mm thickness. The material used for the body of the extrusion die was IMPAX steel. The extrusion die was assembled in five stages. Progressive views of the assembly process are presented in **Figure 2**.

A 3D view of the die along with the screw type used is given in **Figure 3**.

Firstly, the spider head was combined with the male middle mandrel. Subsequently, the outer mandrel and the torpedo were placed in the initial assembly. In stage 3, the die housing was added, and in the following stage (stage 4), the middle ring was adapted to the back side of the spider head. In the last stage (stage 5), the die ring was combined with the middle ring of the previous stage.

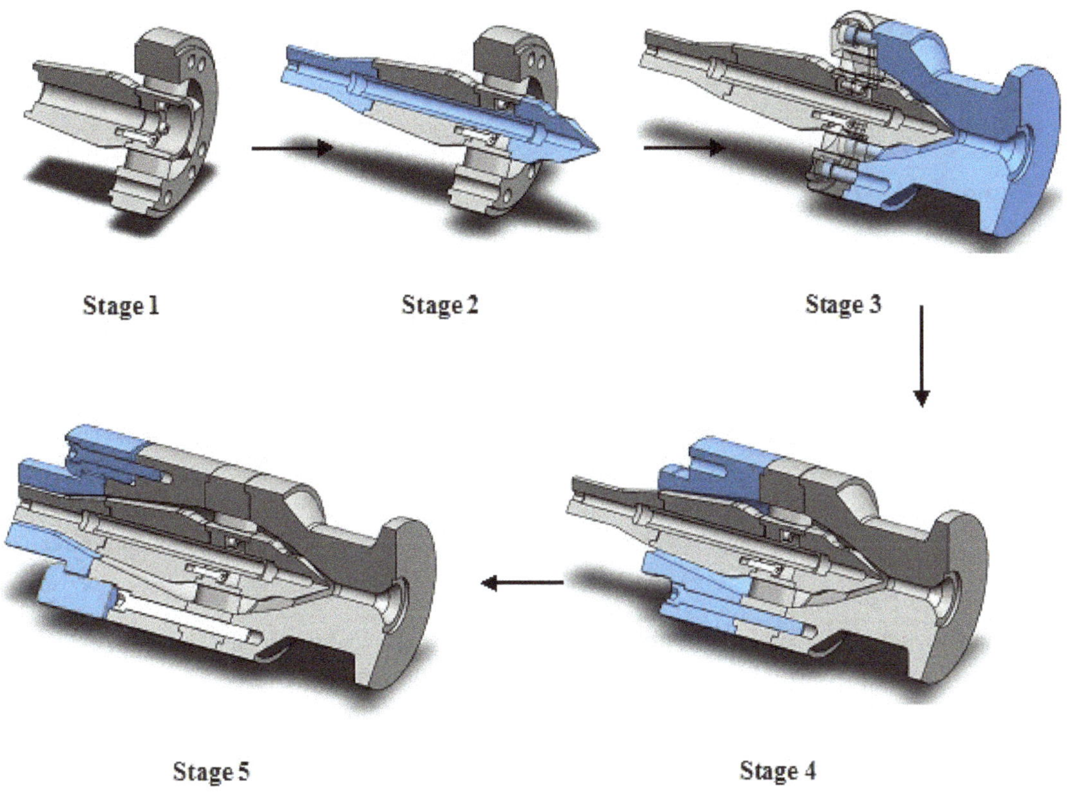

Figure 2. Progressive view of the complete assembly process.

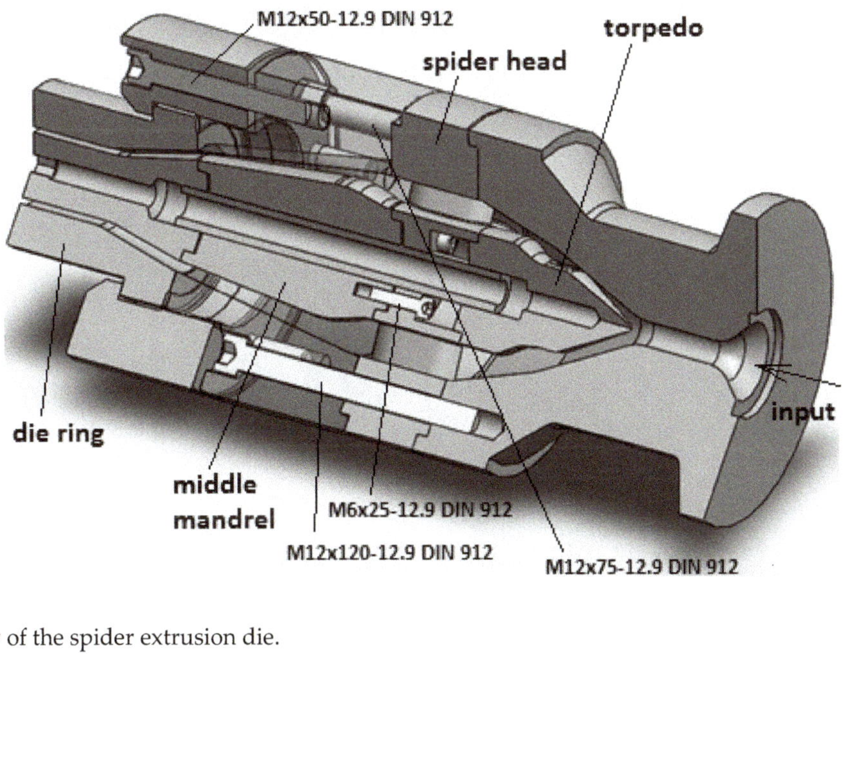

Figure 3. 3D view of the spider extrusion die.

5. Finite element analysis

5.1. CFD analysis

A three-dimensional conjugate heat transfer model, which has been developed for non-Newtonian materials, was processed in the extrusion die. For the numerical solution, the following consideration had been made: a homogeneous and isotropic high-density polyethylene (HDPE) melt with a uniform temperature of T = 469 K is flowing into the spider die. The temperature of the die surface was kept constant at the value T_w = 469 K, and the volumetric flow rate of the polymer melt was fixed at $\dot{V}_{max} = 7.9 \times 10^{-6} m^3/s$.

In most polymer processes, the elastic memory effects can hardly be observed and, therefore, it can be ignored. Since this chapter is concentrated on a qualitative analysis of the flow regimes, the inelastic model was selected as the most appropriate in terms of describing the melt flow.

Due to the polymer melts flow characteristics when it takes place in an extrusion die channel while in steady state, the following assumptions have been made:

- Incompressible steady laminar flow.
- Since the Reynolds number of the melts' flow is extremely low, inertial and gravitational forces are neglected.
- No slipping on the wall interface.
- Uniform and constant die temperature, equal to 469 K.
- Constant volumetric flow rate, equal to 7.9×10^{-6} m^3/s.

The inlet (input), the outlet (output) and the die wall are presented in **Figure 4**.

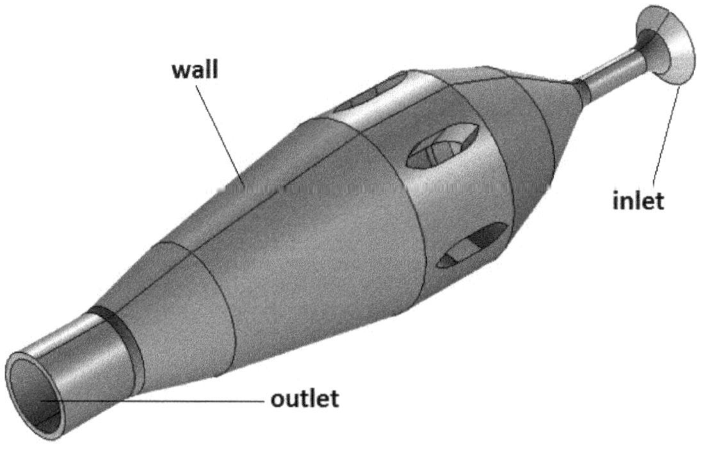

Figure 4. Inner die model used in numerical analysis.

Since the polymer melts are non-Newtonian fluids, the Carreau-Yasuda model was selected to describe dependence of the viscosity on the shear rate and temperature [3]. This model is presented in Eq. (8).

$$n = a_T \cdot n_0(T_R) \left[1 + \left(a_T \cdot \dot{\lambda}(T_R)\, \dot{\gamma}\right)^{\alpha}\right]^{(n-4)/\alpha} \tag{8}$$

where a_T is the shift factor and n_0, λ, α and n are model's fitting parameters.

If ∇ is the Hamilton differential operator, u is the velocity vector, T is the temperature, C_p is the heat capacity and Q is the total source term, the governing equations of the model used can be written in the form below [2, 3]:

$$\text{Continuity equation: } \nabla u = 0 \tag{9}$$

$$\text{Motion equation: } \nabla \sigma = 0 \tag{10}$$

$$\text{Energy equation: } \rho \cdot C_p \cdot u \cdot T = -\nabla q + Q \tag{11}$$

The Cauchy stress vector is given in Eq. (12).

$$\sigma = -p \cdot I + S \tag{12}$$

where p, S and I are the hydrostatic pressure, the extra stress tensor and the Kronecker delta, respectively.

The CFD code of COMSOL 4.3b, using Carreau-Yasuda viscosity model, was used to solve the governing equations. For this model, the effect of the viscous dissipation, that is, the shear heating effect, which is responsible for the fluid temperature increase, was taken into account. This is quite important in polymer extrusion processes and their design.

In order to create the fluid domain, the flow simulation add-in of SolidWorks was used and two lids were created, one in the inlet and another in the outlet of extrusion die as shown in **Figure 5(b)** and **(c)**. Then, the fluid body assembly was created, choosing all the parts of the extrusion die as shown in **Figure 5(d)**. Finally, after deleting all the unneeded subparts, the fluid domain was taken as shown in **Figure 5(e)**.

The mesh model used for the simulations is presented in **Figure 5(f)**. This model included 95,215 tetrahedral finite elements, and the minimum and maximum element sizes were 5.56×10^{-4} and 13×10^{-3} mm, respectively. This mesh was created using automated unstructured mesh generator.

The finite element analysis results for the temperature distribution are presented in **Figure 6(a)**. These results were obtained using the energy balance equations in different positions of the die domain. The input temperature, that is, the temperature of the polymer melt when it enters the die domain, was 469 K. Due to the viscous dissipation, this temperature progressively

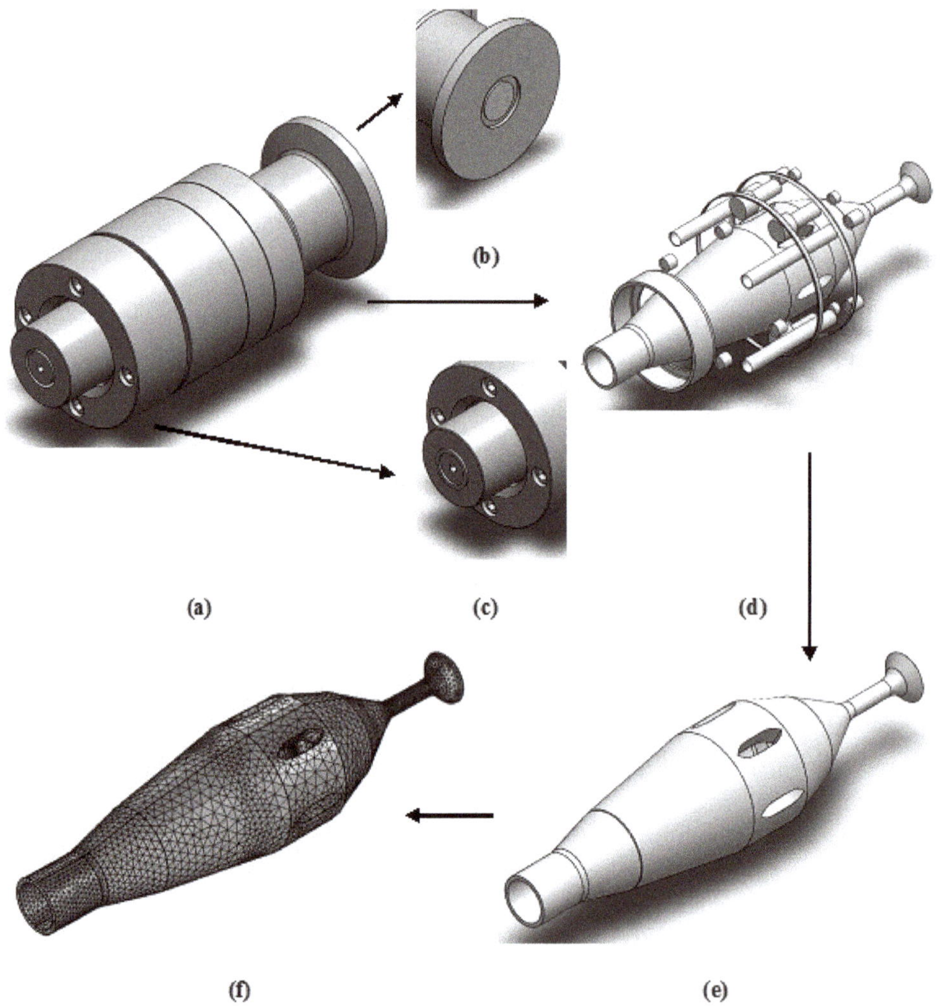

Figure 5. Steps for the geometrical model design (a)–(e) and mesh model of the domain (f).

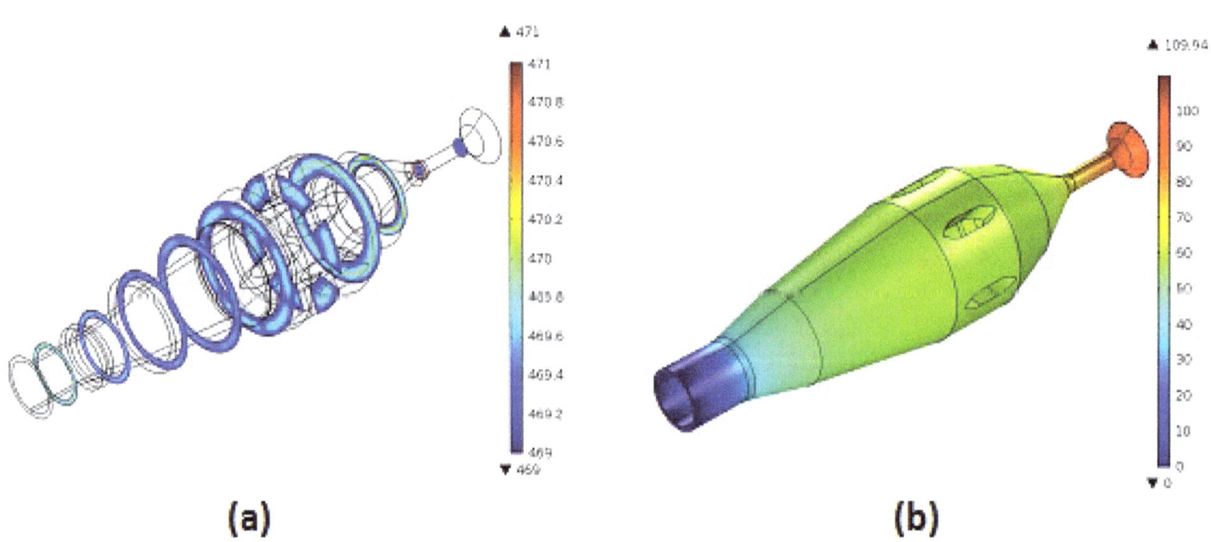

Figure 6. Finite element analysis results for the die domain.

increases, while the flow is moving toward the outlet. The average temperature of the polymer melt is given by Eq. (13) [2], and its increase calculated is equal to 1.53 K.

$$T = \frac{\int_s T \cdot u \cdot ds}{\int_s u \cdot ds} \tag{13}$$

The respective value for the temperature increase calculated using the mathematical model, that is, 4.72 K, was considerably greater in comparison with the finite element analysis results. This is due to the simplification assumption used for the mathematical analysis which indicated that the die walls are adiabatic. In practice, this consideration of the adiabatic wall facilitates the solving process of the mathematical model, but it considerably affects the temperature increase value.

The results of the finite element analysis performed as regards the pressure distribution on the die domain are presented in **Figure 6(b)**. It is obvious that the pressure follows a reduction trend while moving from the extrusion die inlet to its outlet. The pressure drop calculated using finite element analysis was 97.24 bar. This value was similar to the one calculated using the mathematical model, that is, 92.17 bar (Eq. (6)).

5.2. Fluid-structure interaction analysis of the spider head

The most crucial parts of such an extrusion die, as regards its failure under pressure, are the spider legs. The so-called parts of the spider head are the thickest parts of the whole structure, and consequently, these are the most possible points for failure onset under pressure. Therefore, if a single spider leg is able to perform under a specific pressure without failure, it is safe for the whole structure to perform under this pressure.

In order to investigate if the spider legs are able to perform under the pressure occurred using the maximum flow rate provided by the single-screw extruder of this study, a stress analysis was conducted on a single spider leg. This fluid-structure interaction (FSI) problem was solved using the COMSOL Multiphysics software. The solid mechanics continuum equations (Eq. (14)), together with the fluid mechanics Navier-Stokes equations (Eq. (15), were solved using the arbitrary Lagrangian-Eulerian (ALE) method. The deforming geometry dynamics were applied on the boundaries of the moving grid (mesh), and new mesh coordinates were calculated on the channel for each moving step of the boundaries. These moving mesh coordinates were applied on the Navier-Stokes equations. Fixed coordinates were used for the structural parts of the model, that is, for the nonfluid parts, since the strain of these parts was calculated by the COMSOL Multiphysics using structural mechanics. The calculation of the deformed coordinates using ALE formulation was based on the calculated strain of the structural parts.

$$-\nabla \cdot u = 0 \tag{14}$$

$$\rho \frac{\partial u}{\partial t} - \nabla \cdot \left[-p \cdot I + \eta (\nabla u + (\nabla u)^T) \right] + \rho (u \cdot \nabla) u = F \tag{15}$$

In Eq. (15), I is the unit diagonal matrix and F is the volume force which affects the fluid.

Eq. (15), that is, Navier-Stokes, if solved for the velocity u and the pressure p, describes the fluid flow in the channel. Gravitation is not taken under consideration in this model. The same goes for the volume force which affects the fluid. Therefore, the value of the force F in Eq. (15) is equal to zero ($F = 0$). These equations are applied on the deformed coordinates.

The fluid flow at the inlet is described by a fully developed laminar flow and at the outlet is described by a noncompressible flow ($p = 0$). No-slipping conditions, that is, $u = 0$, were applied on the rest of the boundaries.

An elastic and nonlinear model was used in order to apply the large displacement method on the structural domain. Fixed support was applied on the lower and upper spider head boundaries (ring geometry), which indicates an ability lack for movement toward any direction.

The spider head examined is presented in **Figure 7**. In the same figure, the meshing types for the spider leg analysis can be found. The mesh of a single spider leg was unstructured, and it was constituted by 8732 tetrahedral elements. The minimum and maximum element sizes were 1.01×10^{-4} and 2.36×10^{-3} mm, respectively. The flow mesh was also unstructured and was constituted by 62,412 tetrahedral elements. The minimum and maximum element sizes for this type of mesh were 1.17×10^{-4} and 2.03×10^{-3} mm, respectively.

After the mesh-generating process, the solution of the numerical model took place. **Figure 8** presents the finite element analysis results for equivalent stress and total displacement. As can be observed in **Figure 8(a)**, the maximum flow rate of the extruder used leads to the development of a stress equal to 14.79×10^{-2} MPa, which is the maximum stress that can be achieved for HDPE flow with the specific extruder. On the other hand, the yield stress of the IMPAX tool steel used for the die parts is 8×10^{2} MPa. Since the maximum Von Misses equivalent

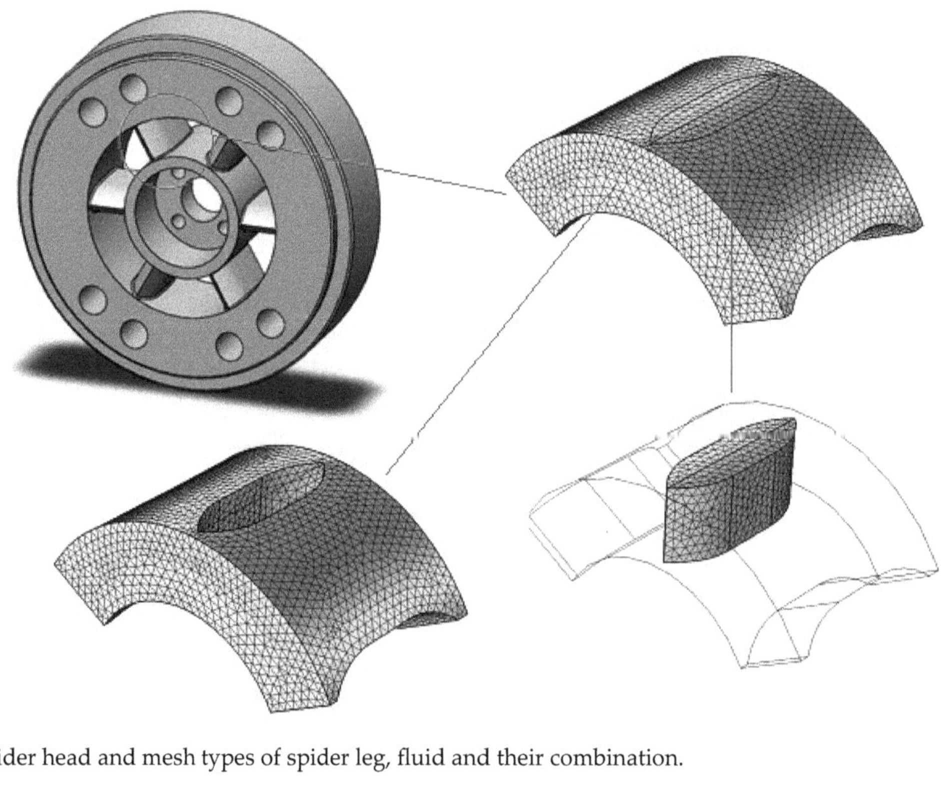

Figure 7. Spider head and mesh types of spider leg, fluid and their combination.

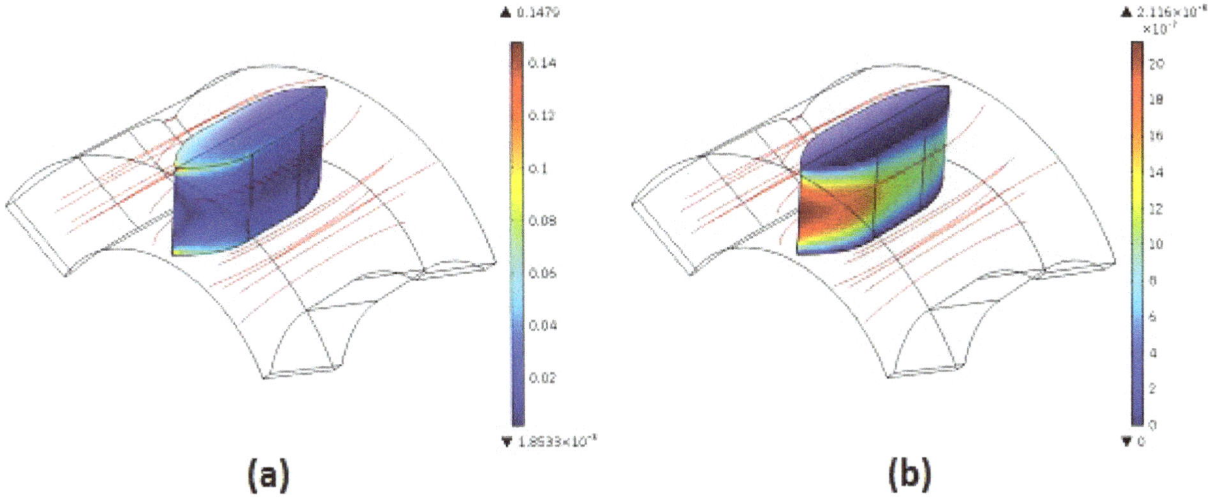

Figure 8. Von misses equivalent stress (a) and total displacement (b) developed on a single spider leg of the extrusion die's spider head.

stress is considerably lower than the yield stress of the spider leg material, the designed spider head, and consequently the entire die, is effective for extrusion processes with the specific single-screw extruder.

6. Manufacturing

The G-code used for the cutting processes applied for the production of each part was generated using SolidCAM software (CAM package). It was transmitted to the CNC cutting

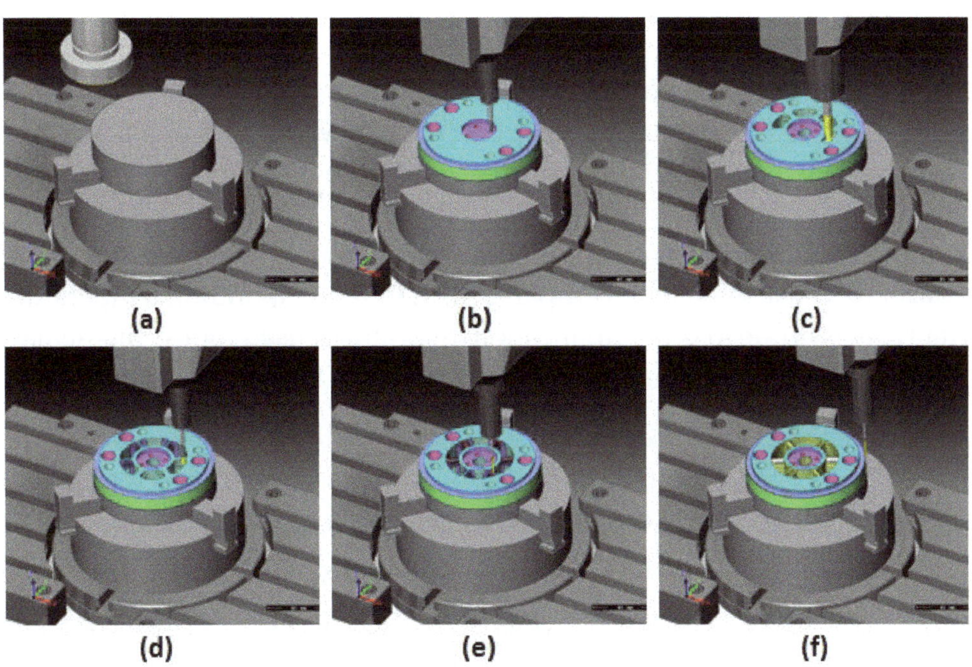

Figure 9. 3D simulation of the cutting process for the production of the spider head in SoliCAM environment.

Figure 10. Johnson Plastics single-screw extruder together with the mounted extrusion die.

machines used, an OKUMA MX-45VE CNC milling cutting center and an OKUMA LB10II CNC lathe, with DNC technology. The 3D simulation of the cutting process for the production of the spider head in SoliCAM environment is presented in **Figure 9**.

Since the spider die was intended to be used for the production of HDPE tubes, it was mounted on a single-screw Johnson Plastics extruder with characteristics: length/diameter ratio 24.1, screw diameter 38 mm and compression ratio 2.75. The extruder with the mounted extrusion die is presented in **Figure 10**.

7. Results and discussion

A summary of the analytical, experimental and numerical results reported in the current chapter including the pressure drop and temperature rise in the die is presented in **Table 1**. The calculated by the mathematical model pressure drop was approximately 3.3% lower than the actual (experimental) result [2]. This can be explained by the fact that the analytical model simplifies the pressure drop calculation in cases of complex geometries. However, the divergence of the calculated value is quite low. On the other hand, the maximum average temperature rise data show 174% divergence compared with the corresponding temperature obtained experimentally. This is due to the simplification assumption used for the mathematical analysis which indicated that the die walls are adiabatic.

The comparison between the experimental and non-Newtonian die flow simulations seems to reveal the expected good agreement with the overall pressure drop as well as with the constant wall temperature boundary conditions.

The pressure data calculated by the numerical Carreau-Yasuda model (97.24 bar) show a fairly good agreement with the experimental results (95.15 bar), whereas the average temperature rise of the molten HDPE was T = 1.53 K, which is approximately 11% higher than the experimental temperature value. This was the value calculated for the boundary conditions

	Mathematical model	Carreau-Yasuda (Numerical) Constant wall temperature	Carreau-Yasuda (Numerical) No heat flux	Experimental
dP (bar)	92.17	97.24	97.24	95.15
dT (K)	4.72	1.53	6.66	1.72
Error dP (%)	3.13	2.19	2.19	
Error dT (%)	174.42	11.05	287.21	

Table 1. Analytical, numerical and experimental data.

of wall with constant temperature. In the case of adiabatic boundary conditions, the average temperature rise of the molten HDPE was considerably higher than the experimental one.

Figure 11 presents the pressure drop throughout all the die zones and the total pressure drop (expressed as percentage) calculated using the mathematical model (**Figure 11(a, b)**) and finite element analysis (**Figure 11(c, d)**). It is obvious that the majority of the pressure drop is observed along the zone V, at the exit of extrusion die.

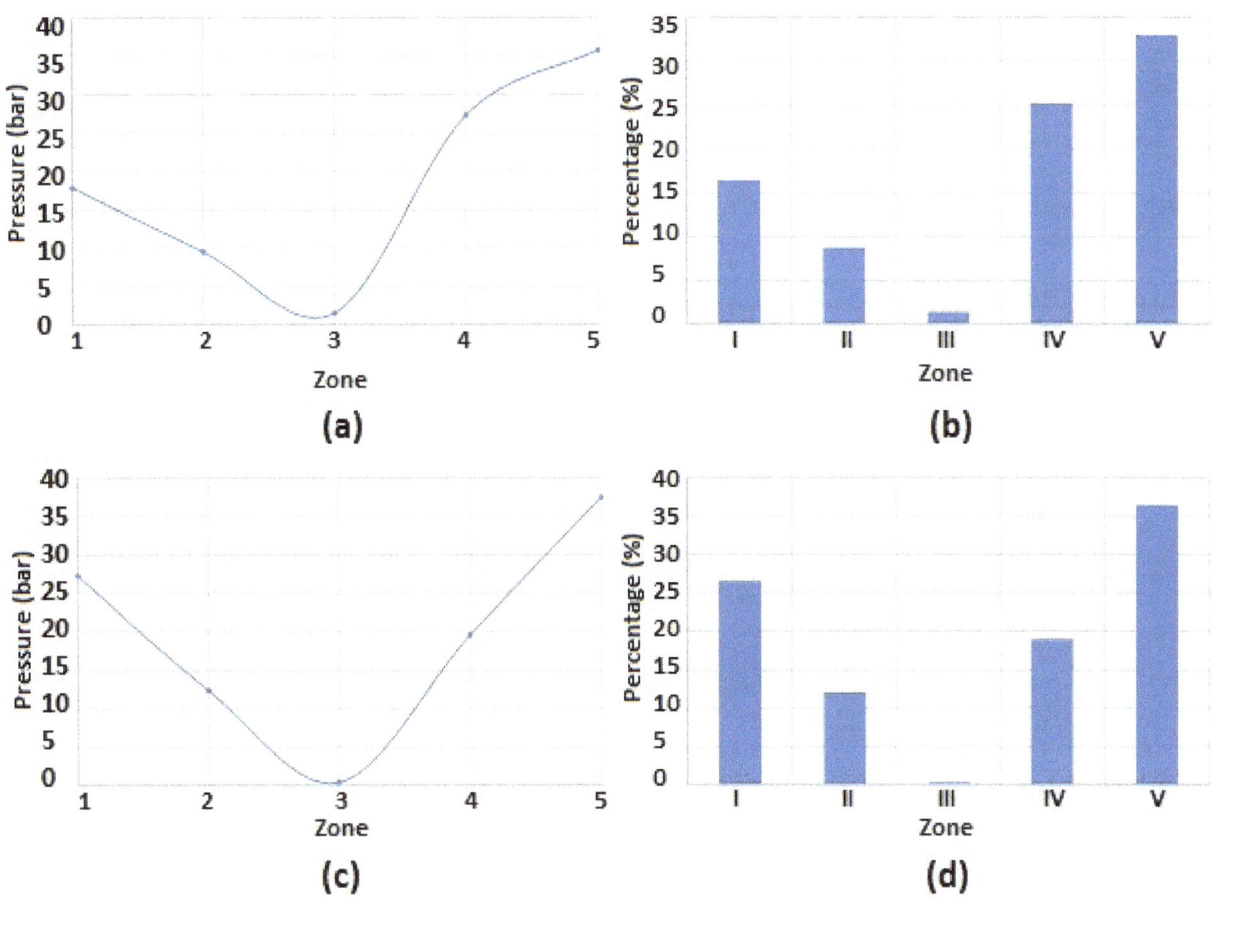

Figure 11. Pressure drop throughout the die zones and total pressure drop calculated using the mathematical model (a, b) and finite element analysis (c, d).

The calculated, using finite element analysis, Von Misses equivalent stresses are significantly lower than the yield stress of the die material, and therefore, it may be concluded that the abovementioned tool steel is suitable for manufacturing spider dies for polymer extrusion applications.

8. Conclusions

Summarizing the main features of the results reported, it may be concluded that there is a significant difference comparing the numerical and analytical models with regard to the temperature developed in the fluid during the extrusion process. This can be explained by the fact that the analytical model is based on the assumption that adiabatic conditions occur, which means that there is no heat transfer between the wall and the polymer material as described above. Finally, it was demonstrated by the stress analysis that the die construction is strong enough to withstand the pressure developed during the die operation and that the stresses do not exceed the material yield strength in any case.

Author details

G.N. Kouzilos*, G.V. Seretis, C.G. Provatidis and D.E. Manolakos

*Address all correspondence to: gkouzilosb@yahoo.com

National Technical University of Athens, School of Mechanical Engineering, Athens, Greece

References

[1] Michaeli W. Extrusion Dies for Plastics and Rubber: Design and Engineering Computations. 3rd ed. Munich: Carl Hanser Verlag GmbH & Co. KG; 2003. p. 362. DOI: 10.3139/9783446401815

[2] Kouzilos GN, Markopoulos AP, Manolakos DE. Manufacturing and modeling of an extrusion die spider head for the production of HDPE tubes. Journal of Manufacturing Technology Research. 2015;**6**(1-2):1-15

[3] Mamalis AG, Kouzilos G, Vortselas AK. Design feature sensitivity analysis in a numerical model of an extrusion spider die. Journal of Applied Polymer Science. 2011;**122**(6):3537-3543. DOI: 10.1002/app.34762

[4] Choudhary MK, Kulkarni JA. Modeling of three-dimensional flow and heat transfer in polystyrene foam extrusion dies. Polymer Engineering & Science. 2008;**48**(6):1177-1182. DOI: 10.1002/pen.20990

[5] Lebaala N, Schmidtb F, Puissanta S. Design and optimization of three-dimensional extrusion dies, using constraint optimization algorithm. Finite Elements in Analysis and Design. 2009;**45**(5):333-340. DOI: 10.1016/j.finel.2008.10.008

[6] Huang GQ, Huang HX. Optimizing Parison thickness for extrusion blow molding by hybrid method. Journal of Materials Processing Technology. 2007;**182**(1-3):512-518. DOI: 10.1016/j.jmatprotec.2006.09.015

[7] Yilmaz O, Gunes H, Kirkkopru K. Optimization of a profile extrusion die for flow balance. Fibers and Polymers. 2014;**15**(4):753-761. DOI: 10.1007/s12221-014-0753-3

[8] Gonçalves ND, Teixeira P, Ferrás LL, Afonso AM, Nóbrega JM, Carneiro OS. Design and optimization of an extrusion die for the production of wood–plastic composite profiles. Polymer Engineering & Science. 2015;**55**(8):1849-1855. DOI: 10.1002/pen.24024

[9] Mamalis AG, Vortselas AK, Kouzilos G. Tube extrusion of polymeric materials: Optimization of the processing parameters. Journal of Applied Polymer Science. 2012;**126**(1):186-193. DOI: 10.1002/app.36555

[10] Rauwendaal C. Polymer Extrusion. 5th ed. Munich: Carl Hanser Verlag GmbH & Co. KG; 2014. p. 950. DOI: 10.3139/9781569905395

4

Changes in Nutritional Properties and Bioactive Compounds in Cereals During Extrusion Cooking

Cuauhtémoc Reyes Moreno,
Perla C. Reyes Fernández,
Edith O. Cuevas Rodríguez, Jorge Milán Carrillo and
Saraid Mora Rochín

Abstract

Maintaining and improving the nutritional quality of foods during processing are the main market and industry concerns. Thus, research should focus on novel and sustainable ways for selecting the appropriate processing method that either increases or does not affect the nutrient content of foods. Thermal processing techniques such as extrusion cooking are widely used for producing breakfast cereals, snack foods, pasta, pet food, etc. Extrusion cooking is a continuous process that uses a combination of high-temperature, high-pressure, and high shear conditions in a short period of time, which results in molecular transformation and chemical reactions within the extruded products. Extrusion cooking brings on many biochemical changes such as denaturation of proteins, gelatinization of starch, lipid modifications, inactivation of microorganisms and enzymes, formation of volatile flavor components, and increase in soluble dietary fiber. Furthermore, extrusion cooking has the potential to improve the nutritional quality of the products by improving starch and protein digestibility and increasing the retention of bioactive compounds with antioxidant properties. Also, this highly efficient technology minimizes water pollution and energy consumption. This review aims to discuss the current information regarding changes in nutritional properties and bioactive compounds in cereals processed by extrusion cooking.

Keywords: extrusion cooking, cereals, nutritional properties, bioactive compounds, food processing

1. Introduction

Improving health through nutrition has been a very demanding and challenging field of study and would continue to be in the future. Maintaining and increasing the nutritional quality of food during processing are a potentially important area for research [1].

Food-processing operations are primarily focused on inactivating disease-causing microorganism (pathogens) and enzymes and on reducing moisture content to concentrate the processed foods. However, it is important to keep in mind that several changes in foodstuffs including appearance, composition, and nutritional and sensorial properties (color, texture, and flavor) can occur during processing. Interestingly, recent research has now established that food-processing operations have positive effects that improve the quality and health benefits of food [2].

Extrusion cooking is one of the most important food-processing alternatives. This process has been used since the mid-1930s for the production of breakfast cereals, ready-to-eat snack foods, and other textured foods. Furthermore, at this moment, extrusion process has become the major processing technology for food and feed industries, and it is rapidly evolving from an art into science and technology [3].

Extrusion cooking is preferred over other food-processing methods because it is a high-temperature short-time (HTST) process, which preserves important nutrients, denatures antinutritional components of foods (trypsin inhibitors, tannins, and phytates), disinfects the final product, and maintains normal colors and flavors of the food [4, 5].

Consumers have developed a growing understanding of how the composition of food products can impact on the nutritional quality of foods. Likewise, bioactive compounds or phytochemicals in food and food products play a relevant role in human health, providing protection against many chronic and degenerative diseases [6].

2. General and nutritional aspects of cereals

2.1. General aspects of cereals

Cultivation of cereal grains was the first agricultural attempt of the early man, and we still enjoy them today depending on the region we live and what grows there well [7]. Botanically, cereals are classified as grasses and belong to the monocot family Poaceae, also known as Gramineae. Within each cereal species, numerous varieties produced by breeding exist in order to optimize agronomical, technological, and nutritional properties [8]. The major cereals consumed in the world include wheat, corn, rice, oats, rye, barley, buckwheat, sorghum, and millet as minor grains. Wheat, corn, and rice take up the greatest part of the land cultivated by cereals and produce the largest quantities of cereals grains [8].

Cereals are composed of (1) endosperm (the main part of the grain, mainly starch), (2) germ (the smallest part of the grain; it contains vitamin E, folate, thiamine, phosphorus, and magnesium), and (3) bran (the outer layer of the grain that contains fiber omega-3 fatty acids, vitamins, dietary minerals, and phytochemicals). Whole grains are important components of the human diet as shown by inclusion in the Food Guide Pyramid and US Dietary Guidelines [2, 7, 9]. However, cereal grains have poor protein quality due to their low content of the essential amino acids lysine and tryptophan, which are referred to as the main limiting amino acids in cereals [10].

In their natural form (as in whole grain), cereals are a rich source of vitamins, minerals, carbohydrates, fats, protein, and phytochemicals. However, when refined by the removal of the bran and germ, the remaining endosperm is mostly composed of carbohydrates and lacks the majority of the other nutrients [7].

Epidemiological studies have shown that a habitual intake of whole-grain cereal products is inversely associated with the risk for developing various types of chronic diseases such as obesity, cardiovascular disease, Type II diabetes, and some types of cancer [2, 11]. It has been suggested that these health benefits are not attributed only to a single component, but to the combined effects of several of them: dietary fiber, phenolic acids and flavonoid compounds, and other bioactive components present in cereal grains [11, 12].

Phenolic compounds exist in free and bound form in cereals. Bound phenolic compounds are referred to as phenolic compounds and are covalently bound with the cell wall components [13]. Phenolic acids are the most common phenolic compounds in cereals. They have a strong antioxidant activity and may modulate cellular oxidative status and protect biologically important molecules from oxidative damage such as DNA, proteins, and membrane lipids [14].

Cereal products are a complex multicomponent system that might contain mixtures of protein, vitamin, minerals, oils, polysaccharides, and bioactive compounds or phytochemicals. Hence, there is a reason for the increasing demand of new processed foods from whole cereals; they are convenient and nutritious to satisfy the demands of health-conscious consumers.

2.2. Nutritional and bioactive compounds in cereals

Cereal grains are grown in greater quantities and provide more food energy worldwide than any other types of crops. They are therefore staple food crops [7]. About 50% of the calories consumed by the world population originate from three cereals: rice (23%), wheat (17%), and maize (10%) [15, 16]. These are grown for their highly nutritious edible seeds, which are often referred to as grains.

Besides the main energy source, they supply a variety of nutrients and other food components such as bioactive compounds or phytochemicals [13]. Cereals are an important source of calories for humans, both by the direct intake and as the main feed for livestock [13]. Dietitians suggest to include whole-grain products in the diet, which provide large quantities of macronutrients (carbohydrates, proteins, and lipids), vitamins (particularly of the B group and vitamin E), and micronutrients, such as selenium, zinc, copper, and magnesium [17, 18]. Besides, cereals contain phytochemicals like phenols, phytoestrogens, and fermentable carbohydrates such as dietary fiber and resistant starch or oligosaccharides, which have been recently associated with cholesterol lowering, cardiovascular disease protection, and cancer-risk decrease [18, 19].

The chemical or nutritional composition of cereal grains is mainly characterized by the high content of carbohydrates. The carbohydrates for the human diet are classified in bioavailable and not bioavailable (or fiber). The bioavailable fraction, mainly starch, is deposited in the endosperm (comprising 65–75% of their total weight). The non-bioavailable fiber located in the bran (2–13%) is constituted by polymers such as arabinoxylans (1.5–8%), β-glucans (0.5–7%), sugars (≈3%), cellulose (≈ 2.5%), and glucofructans (≈ 1%). The other nutritional or chemical compounds are the proteins, which are the second most important compounds of cereals (comprising range of 7–11%). Lipids are located mainly in the germ fraction (average of 2–4%); a small fraction of them are in the aleurone layer and to the lesser extent in the endosperm. Cereal lipids have a similar fatty acid composition in which linoleic acid contents reaches 39–69%, while oleic acid and palmitic acid make up to 11–36 and 18–28%, respectively. Minerals in cereals are the minor constituents (1–3%). Nevertheless, the high content of vitamin B is of nutritional relevance [7, 8, 20]. Depending on the structure of each grain and the amount of their chemical constituents, there exist significant differences among cereals and even among species and varieties within each cereal.

More than 25,000 bioactive food constituents are present in the diet, and cereals are not the exception. The majority of food phytochemicals are also antioxidants, and many of these compounds play a role in modifying processes involved in the development of various diseases. It has been reported that these phytochemicals can also work in vivo as individual compounds, exerting their antioxidant property by neutralizing reactive oxygen or reactive nitrogen compounds and contributing to antioxidant defense of the body, thus promoting longevity, cell maintenance, and DNA repair [21, 22].

The most important groups of bioactive compounds in grains include phenolic compounds (phenolics acids, alkylresorcinols, and flavonoids), carotenoids, vitamin E, dietary fiber, and β-glucan [14, 24]. Phenolic compounds are secondary metabolites of plants and are part of the defense system against the sun's ultraviolet light as well as pathogens. These compounds play an important role in combating oxidative stress in the human body by maintaining a balance between oxidants and antioxidants [23, 24]. The antioxidant properties of cereal grains are mainly attributed to phenolic compounds [25]. Hence, the presence of phenolic compounds in cereals and their distribution can protect against cancer, cardiovascular diseases, diabetes, hypertension, asthma, and even infection if consumed in abundance in cereals [14, 22, 26].

Phenolic compounds (mainly phenolic acids) are concentrated in the bran fraction of cereal grains and exist in free, soluble conjugated, and insoluble bound forms [27]. Ferulic acid is the most abundant phenolic acid in cereal grains, followed by *p*-coumaric, sinapic, and caffeic acids [28]. In wheat and maize, ferulic acid represents up to 90% of the total phenolic acids and <95% in bound form [27, 29]. However, in other cereals such as finger millet (*Eleusine coracana*), phenolic acids represent 71% of the free-form phenolics, being protocatechuic acid the most abundant of them [30].

Cereals and other food items have to be processed before consumption to improve digestion and facilitate metabolism in humans [2]. Various processing technologies have been developed to enable the release and/or to increase the accessibility of nutritional and phytochemical components in cereal grains. Among these, the most common technique in food processing is to apply mechanical treatment resulting in reduction of particle size, breakdown of cereal matrices, or degradation of fiber polymers. Likewise, thermal processing techniques such as steaming, autoclaving (pressured steam heating), drum drying, roasting, and microwave heating [11] are often utilized. Among the thermal treatments, the high-temperature short-time extrusion cooking technology has proven to have limitless applications in processing of cereal-based product.

3. Changes in extrusion cooking of cereals

Nowadays several breakfast cereal products are produced by flaking, oven and gun puffing, baking, shredding, and direct expansion. Extrusion cooking technology is becoming popular over other common processing methods due to its automated control, high capacity, continuous operation, high productivity, versatility, adaptability, energy efficiency, and low cost. Moreover, it also enables design and development of new food product, high product quality, unique product shapes and characteristics, energy saving, and no effluent generation [31, 32]. Extrusion cooking of breakfast cereals is a much easier and cheaper processing method than the conventional ones, and many components may be incorporated in the recipe [18].

There are three major types of extruders being used in the food industry: piston extruders, roller-type extruders, and screw extruders. Screw extruders are the most common extruders used these days and can be categorized as single- and twin-screw extruders [33]. In the single-screw extrusion cooking process, the extruder can be divided into three regions: conveying, swelling, and melting/degradation, in terms of the transition of cereal starch. Both the conveying and swelling regions are located in the cooling zone, where the flow pattern behaves as a plug flow reactor. The melting of starch granules and degradation of starch molecules occur simultaneously in the third region. The flow pattern is changed from plug flow reactor to continuous stirred tank reactor; thus, more mixing and longer residence time occurred in the heating zone. However, with twin-screw extruders, it is common to employ a section of spirally flighted screw elements behind the die head zone to provide a steady pumping action and to generate high die pressure [33].

Extrusion cooking of foods has been practiced for over 50 years. The food extruder, which was initially limited to mixing and forming macaroni and ready-to-eat cereal pellets, is now considered a high-temperature short-time bioreactor that transforms raw ingredients into modified intermediate and finished products [33]. Besides, several food products are developed by extrusion, i.e., pasta, breakfast cereals, bread crumbs, biscuits, crackers, croutons, baby foods, snack foods, confectionery items, chewing gum, texturized vegetable protein, pet foods, dried soups, dry beverage mixes, and instant flours to make tortillas [33–35].

During extrusion cooking, its simultaneous actions of temperature, pressure, and shear, along with their intensities and interactions vary enormously depending on feed ingredients, extruder configuration, and the desired characteristics of the final product [25]. This process consists in using high temperature and short time in which moistened, expansive, and carbohydrate or protein food materials are plasticized and cooked in a tube.

The combination of moisture, pressure, temperature, and mechanical shear results in molecular transformations and chemical reactions that allow obtaining expanded products and porous structure [1, 36, 37]. The expansion is the result of both the elastic swelling and bubble growth effects. The characteristics of the products obtained by extrusion cooking depend on the moisture content, extrusion temperature [38], residence time, pressure, and shear [33, 39].

The extrusion cooking process is preferable to other food-processing techniques in terms of its high productivity and significant nutrient retention due to the high temperatures and short times that are required [4]. Likewise, this technology has several applications including increasing numbers of ready-to-eat cereals, salty and sweet snacks, coextruded snacks, indirect expanded products, croutons for soups and salads, an expanding array of dry pet food and fish foods, textured meat-like materials from defatted high-protein flours, nutritious precooked food mixtures for infant feeding, and instant flour to make tortillas [32, 40, 41]. Moreover, extrusion has been used in the development of expanded cereals that include the addition of other ingredients (i.e., fruits and vegetables) in order to increase the health benefits [42].

Interest has grown in the physicochemical, functional, and nutritionally relevant effects of extrusion processing. Prevention or reduction of nutrient destruction, together with improvements in starch or protein digestibility, is clearly of importance in most extrusion applications [1]. Another advantage of extrusion cooking is the destruction of antinutritional factors, especially trypsin inhibitors, hemagglutinis, tannins, and phytates, all of which inhibit protein digestibility [1, 43]. Several extrusion-processing conditions are accounted for the quality of finished products. The control of feed rate, screw speed, barrel temperature, and barrel pressure, together with the abovementioned critical parameters, will determine the crispness, hardness, and various other characteristics that will influence the success of the final product [44].

3.1. Effect of extrusion process parameters on physical and chemical properties of extruded products

Cereal thermal processes such as baking, roasting, and extrusion cause a number of physical and chemical changes due to starch gelatinization, protein denaturation, component interactions, and browning reactions. These changes result in improved organoleptic properties, increased nutrient availability, and inactivation of heat-labile toxic compounds and enzyme inhibitors [25].

Food products obtained by extrusion technology are composed mainly of cereals, pulses, and/or vegetable proteins. The major role of these ingredients is to give structure, texture, mouth feel, and many other characteristics desired for specific finished products [33].

Functional properties of extruded foods play an important role in their acceptability including water absorption, water solubility, oil absorption indexes, expansion index, bulk density, and viscosity of the dough [33]. Several changes in the matrix food have been reported using the extrusion process, which are the result of the combination of moisture content of the starting materials, pressure-temperature, and screw speed, which are responsible of physical and chemical transformations in the final product and therefore affect product quality [42].

Extrusion cooking technology brings numerous chemical changes, such as gelatinization or starch, denaturation of protein, lipid modification, as well as inactivation of enzymes and microorganisms [45]. Denaturation of grain proteins during extrusion cooking allows the opening of loose structures for tannin-protein interactions causing the formation of tannin-protein complexes and retention of antioxidant activity. These complexes are broken down in the human gastrointestinal tract to release bound tannins and act as free radical scavengers [46]. Another important chemical effect occurring during extrusion is the browning or the formation of Maillard reaction products (MRP), which contributes to the antioxidant properties of the final product [47].

However, several investigators have reported that during extrusion, there is the loss of bioactive compounds with antioxidant properties (i.e., phenolics, tocopherols, carotenoids, anthocyanins, flavonoids, tannin, and other bioactive compounds) [48, 49]. Thermal degradation of phenolic compounds may be due to complex formation with Maillard reaction by-products and high moisture content promoting phenolic polymerization [50] affecting their extractability and antioxidant activity [51]. In contrast, earlier research on cereal products has shown that thermal processing might contribute in releasing bound phenolic acids by breaking down cellular constituents and cell walls [52]. For instance, transformation in more easily extractable forms of phenolic compounds has been reported in single-screw extruders with low moisture contents (<15/100 g), high shear stress, and high temperatures [53]. All of these chemical changes are associated with structural changes that occur in the materials subjected to extrusion increasing the release of the bioactive compounds in the cell wall matrix [54], thus making these materials more easily extractable [55]. All these physical changes are related to high shearing force in combination with high temperature and pressure that can efficiently disintegrate the rigid cell walls of the matrix food.

3.2. Effects of extrusion cooking process on nutritional properties of cereals

Nutritional properties in cereals are provided by two main groups of nutrients: macronutrients comprising of lipids, proteins, and carbohydrates, and micronutrients that include vitamins and minerals. Many researchers have reported both the positive and negative effects of the extrusion process on the nutritional quality of food and feed mixtures. These results are dependent on the different extruder conditions (temperature, feed moisture, screw speed, and screw configuration) and raw material characteristics (composition, particle size) [1].

As discussed above, extrusion cooking has been studied extensively to produce a wide variety of foods [33–35]. Interestingly, as a multistep, multifunction thermal/mechanical process, extrusion could have beneficial or detrimental changes on bioavailability and content of nutrients of cereal products [25]. On the one hand, extrusion (1) induces starch gelatinization improving its digestibility, (2) promotes the destruction of antinutritional factors (undesirable enzymes, trypsin inhibitors, and hemagglutinins), (3) increases the content of soluble fiber, (4) improves protein digestibility, and (5) reduces lipid peroxidation [1]. On the other hand, extrusion can also negatively affect the bioavailability of certain nutrients [56]. The heat-labile vitamins and some amino acids are lost, and the Maillard reaction that occurs during the process can reduce the nutritional value of the proteins [1]. In this section, we will discuss the various effects of extrusion cooking on the nutritional properties of cereals.

3.2.1. Macronutrients

3.2.1.1. Lipids

While lipids can act as lubricants during extrusion, the amount of lipid content can affect the properties of the extrudates. Hence, the presence of lipids in less than 3% does not affect extrusion, but in quantities above 5% can reduce expansion rate, and above 10% they reduce slip within the extruder barrel, making extrusion difficult [57].

Extruded foods, particularly expanded products, are susceptible to lipid oxidation, one of the main causes of food deterioration [58]. Although there is not much research focused on the nutritional changes in lipids after extrusion, it has been reported that extrusion cooking can minimize lipid oxidation, thus increasing the nutritional and sensory quality and shelf life of foods [1]. Among the factors involved in the delay of oxidation in extruded foods are (1) denaturation of lipase and other enzymes that may contribute to oxidation [58]; (2) formation of lipid-amylose complexes, thereby reducing both starch and lipid availability and increasing oxidative stability and shelf life of extruded products [59]; (3) release of endogenous antioxidants in grains during extrusion that may provide protection against peroxidation [60]; and (4) creation of Maillard reaction products with antioxidant activity. In this regard, Sproston et al. [61] recently showed that a Maillard reaction product (MRP) derived from D-glucose and L-cysteine possesses antioxidant properties and that such product could be useful in inhibiting lipid oxidation in complex emulsions.

Polyunsaturated fatty acids (PUFA) ω-3 and ω-6 are essential for humans, and because of their greater number of instaurations, they are particularly sensitive to oxidation [62]. Thus, food-processing alternatives that result in minimal loss and the lower degree of oxidation of these components are desirable. Suzuki et al. [63] demonstrated that PUFA ω-3 and ω-6 in salmon muscle were retained after extrusion and its pretreatment, suggesting that foods rich in PUFA can be processed through extrusion without significant losses. In contrast, Ramos Diaz et al. [64] studied the effect of extrusion on corn-base snacks containing different combinations of amaranth and quinoa. They reported that extrusion substantially reduced the content of fatty acids (palmitic, oleic, linoleic, and linolenic acids) and α-, β-, and γ-tocopherols compared to flour blends. The authors attributed the considerable reduction in the content of fatty acids and tocopherols during extrusion to the formation of amylose-lipid complexes [59, 65]. It is also important to note that in the above-discussed studies, the ingredients used were different; hence, the available amylose for lipid binding and the formation of complexes was also different [59].

3.2.1.2. Proteins and amino acids

Cereals (maize, sorghum, rice, barley) and pulses (beans, peas, chickpeas, lentils, and other dry edible seeds) have traditionally been the dominant dietary plant protein source [66]. Protein nutritional value depends on the content of essential amino acids and the digestibility and utilization of the protein [67]. Several factors can affect protein digestibility of cereals, among them are the grain structure and composition, the presence of disulfide bonds, surface functional groups, and protein hydrophobicity and conformation [68]. Also, processing such as pressure, temperature, fermentation, freeze/thaw cycles, and shear can also modify protein digestibility [69].

Several reports have shown that extrusion can improve protein digestibility by denaturing proteins and exposing of enzyme-susceptible sites. This phenomenon is attributed to the effects of high shear on protein structure and conformation that occur during extrusion, leading to the manufacture of products with highly digestible proteins [70]. Vaz and Arêas [70] showed that the increase in protein solubility observed in extruded meat-based formulations was associated with protein degradation and denaturation during the process. Similarly, enzymes and enzyme inhibitors generally lose activity during extrusion unless they are stable to heat and shear. Reductions in protease inhibitors can contribute to better plant protein utilization [57].

In addition, extrusion has been proposed as a viable alternative to influence allergenic properties of food proteins. The potential reduction in antigenicity is due to degradation of protein structures that ultimately results in the reduction of IgE- and IgG-binding capacity during thermal processing of foodstuffs [71].

3.2.1.2.1. Maillard reaction and advanced glycation end products

As it has been discussed, extrusion process under high pressure causes major chemical changes, thermal degradation, dehydration, depolarization, and recombination of fragments, all of which can promote glycoxidation [69, 72]. The concentration of advanced glycation end products (AGE) in foods, which are formed by Maillard reaction, has been demonstrated as a risk factor associated with the etiology of age-related diseases in humans, such as atherosclerosis, nephropathy, retinopathy, osteoarthritis, neurodegenerative diseases, and diabetes mellitus [73]. In addition, AGE by binding to their receptors (RAGE), which are found in a wide variety of cells, can lead to oxidative stress, vasoconstriction, and inflammatory responses. The AGE can covalently cross-link tissue proteins and, thereby, modify structural and functional properties of the proteins [74, 75]. During extrusion, the Maillard reaction is sometimes induced to contribute to desired flavor and color and to enhance palatability [76]. However, excessive Maillard browning can result in losses of lysine, destruction of vitamins, and reduction of bioavailability of trace elements [77]. Retention of lysine in the breakfast cereals is considered most important since it is the limiting amino acid among most of cereal snacks [10]. Thus, the Maillard reaction can result in unfavorable consequences such as a decreased protein quality due to the loss of bioavailable essential amino acids and, as mentioned before, the production of AGE. Future studies should focus on the optimization of processing conditions in a way that the desired beneficial effects are promoted, and the undesired effects are minimized.

3.2.1.3. Carbohydrates, starch, and fiber

One of the more widely researched aspects of extrusion on the nutritional content of products is the way extrusion technology can affect carbohydrate digestibility. Starch is usually the major food constituent in extruded foods such as breakfast cereals, snacks, and weaning foods. Humans do not readily digest raw starch [78]. However, the digestibility of starch may be improved by the extrusion process due to its partial gelatinization and fragmentation attributed to the effect of shear and temperature. The depolymerization of the starch allows it to be more readily available to digestive enzymes. Moreover, during extrusion the physical breakdown of starch molecules takes place, resulting in smaller and more digestible fragments [79]. The extrusion process can increase the available digestible carbohydrate in cereals by up to threefold compared to raw (unextruded) cereals [80]. In the literature, there are several examples illustrating that extrusion improves starch digestibility.

Borejszo and Khan [81] found that sucrose, raffinose, and stachyose decreased significantly in extruded pinto bean starch fractions. While Alonso et al. [82] reported that compared to raw peas, starch and stachyose were lower in extruded peas [57]. Similarly, Mahasukhonthachat et al. [83] reported that the rate of starch digestion of sorghum increased by tenfold after extrusion when compared with non-extrudates, while Haralampu [84] reported that 22% of the resistant starch was lost (i.e., increase in more digestible starch), possibly due to high shear.

High starch digestibility is desirable in the food industry for the manufacture of specialized nutritional foods such as infant and weaning foods or to target particular consumer needs (elderly requiring rapidly digestible forms of starch, people participating in athletic activities, and those looking to reduce the content of indigestible oligosaccharides that cause flatulence in foodstuffs). However, these products tend to induce a higher glycemic response than their unprocessed raw ingredients. High blood levels of postprandial glucose and insulin have been implicated in the development of insulin insensitivity and chronic metabolic diseases such as Type II diabetes and cardiovascular disease [85]. By altering not only the digestibility of starch but also the conformation of starch, extrusion offers the ability to reduce the high glycemic index of some foods by converting starch to digestion-resistant starch. Hence, the formation of resistant starch by extrusion may have value to promote reduced calories in food products [86]. In this respect, an interesting observation is that extrusion can also increase the amount of resistant starch and soluble dietary fiber present in extrudates.

Different researchers have reported an increase in enzyme-resistant starch content in wheat, maize [87], and barley (2–3%) after extrusion [88]. In regard with the fiber content, Jing et al. [89] optimized extrusion process parameters (temperature 115°C, feed moisture 31%, and screw speed, 180 rpm) to obtain the highest values of soluble dietary fiber in soybean residues. Under these conditions, the soluble dietary fiber content residue increased by 10.6% and had higher water retention, oil retention, and swelling capacities than unextruded residues. Similar results were found by Chen et al. [90] using optimal conditions (170°C and an extrusion screw speed of 150 rpm/min) for blasting extrusion in soybean residue. These researchers found that the content of soluble dietary fiber from soybean residues increased by more than tenfold and showed improved water solubility, water retention capacity, and swelling capacity compared to unprocessed soybean. Furthermore, they tested the physiological effects of their high dietary fiber product and observed that it was able to significantly reduce total cholesterol, low-density lipoprotein cholesterol, and triglyceride levels in vivo.

The increase in soluble dietary fiber in extruded products could be explained by the formation of additional components by transglucosidation, whereby 1,4 carbon-oxygen bonds are cleaved, and new anhydroglucose linkages are formed, and the resulting novel bonds would be resistant to digestion by enzymes [91]. Another possibility is to increase insoluble dietary fiber with the formation of retrograded amylose, insoluble at room temperature [91, 92]. This could also be attributed to the formation of covalent interactions between macronutrients leading to components that are insoluble and not hydrolyzed by digestive enzymes [92, 93]. These indigestible glucans may be Maillard reaction products likely resulting from chemical reactions between starch and proteins present within the dietary fiber-containing matrix [93].

3.2.2. Micronutrients

3.2.2.1. Vitamins

In general, vitamins differ greatly in chemical structure and composition. Their stability during thermal process is also variable. The extent of degradation depends on various parameters during food processing and storage, e.g., moisture, temperature, light, oxygen, time, and pH [72]. Extrusion cooking has a significant effect on the stability of hydrosoluble vitamins. For instance, higher barrel temperature and low feed moisture induce ascorbic acid degradation during extrusion [6]. On the other hand, using short barrel (90 mm) extruders result in higher retention rate of the vitamin B group (44–62%) than long barrel extruders (20%). The retention of vitamins is generally not related with their initial level in foods and varies with cereal type. In corn, pyridoxine (vitamin B6) showed stability to extrusion cooking compared to oats and corn/pea ingredients [94]. In another study of fortified corn extrudates, the effect of temperature during the extrusion process on thiamine (B1) and riboflavin (B2) content was explained [95]. The authors found no significant differences between riboflavin content in traditional extrudates produced at feed moistures of 80 and 110°C barrel temperature. Interestingly, riboflavin content of extruded products at 20% feed moisture was higher than the one produced at 25% feed moisture at 130°C.

Extrusion cooking also affects the stability of fat-soluble vitamins such as vitamins A and E [96], which are natural antioxidants in cereal grains. The levels of vitamin E decreased (63%) in extruded buckwheat. Likewise, in other extruded cereals (oat, barley, wheat, rye, and buckwheat), a significant decrease (63–94%) in tocopherols and tocotrienols was observed [55, 97]. Isomers of vitamin E such as α-tocopherol and α-tocotrienol are the least resistant to temperature compared to other isomers [97].

3.2.2.2. Minerals

Minerals are considered stable during heat treatment. However, smaller molecules may be affected by either the extrusion process itself or changes in larger molecules, which in turn can affect other compounds present in the food [98]. To date, few studies have reported mineral stability during extrusion cooking of cereal grains. Interestingly, extrusion cooking can improve the absorption of minerals by reducing other factors that inhibit their absorption [99]. Phytates may form insoluble complexes with minerals decreasing their bioavailability in the gastrointestinal tract. However, extrusion may hydrolyze the complex phytate minerals to release phosphate molecules [1].

In a study, a range of 13–35% reduction in phytate content from a wheat bran-starch-gluten extruded mix was reported in Ref. [100]. Polyphenols may also inhibit mineral absorption. The presence of tannins can form insoluble complexes with divalent ions in the gastrointestinal tract, inhibiting their bioavailability. Interestingly, extruded foods could also exhibit an increase in mineral absorption, which may be attributed to the destruction of polyphenols during the extrusion cooking [1, 99]. Similarly, absorption of minerals could be impaired by fiber (components such as cellulose, lignin, and hemicelluloses). However, the high temperature during the extrusion could reorganize the fiber components modifying their chelating properties [82].

3.3. Effects of extrusion cooking process on bioactive compounds of cereals

3.3.1. Phenolic compounds (phenolic acids and anthocyanins)

Aside from their nutritional contribution, cereals contain bioactive compounds with potential health benefits. These biologically active phytochemicals are found to be natural antioxidants and could be beneficial in reducing the risk of many diseases [1]. Food processing results in the destruction or change of natural bioactive compounds, which may affect the antioxidant properties of foods [2, 101]. According to the literature, food processing can alter antioxidant activity positively and negatively. In this regard, extrusion cooking can have either effect on the phenolic content in cereal grains.

On the one hand, extrusion causes decomposition of heat-labile phenolic compounds and polymerization of some others [102], resulting in the decrease of the extractable phenolic content. On the other hand, extrusion disrupts cell wall matrices and breaks covalent bonds in high-molecular-weight polyphenol complexes [53], improving the phenolic accessibility. The net effect of extrusion on total phenolic content depends on which of these phenomena are predominant [11]. Furthermore, antioxidant activity is correlated with the presence of bioactive compounds such as phenolics, carotenoids, flavonoids, and anthocyanins in foods [103].

Recently, several studies have reported the effect of extrusion and extrusion conditions on the phytochemical content and antioxidant activity of cereal grains. Important losses or increases of bioactive compounds are reported due to the thermal effects and chemical changes that occur during extrusion [42]. For instance, the release of phenolic compounds is highly dependent on moisture content, time, and temperature [25, 104].

In a study on dark buck wheat flour, no change in antioxidant capacity after its extrusion at 170°C was reported in Ref. [105]. Another study showed a significant reduction in both antioxidant capacity (60–68%) and total phenolics (46–60%) in barley extrudates compared with unprocessed barley flour [102]. Zielinski et al. [97] found significant changes in selected cereals (wheat, barley, rye, and oat) during extrusion cooking at different temperatures (120, 160, 200°C). The authors found significant increases in phenolic acids (mainly ferulic acid), while sinapic and caffeic acids were not detected in the extruded grains. Other authors are using an optimized technique to obtain extruded products (cereals or mixture cereal/legume) with high nutritional quality and high antioxidant value. A ready-to-eat expanded snack with high nutritional and antioxidant value was developed from a mixture (70/30) of whole amaranthine, transgenic maize, and black beans by optimizing the extrusion process [106]. Using optimal conditions for extrusion, these authors found an increase in total phenolic content (74%) and antioxidant activity evaluated as ORAC (18%) and ABTS (20%) in the extruded snack with respect to the unprocessed whole-grain mixture. In general, the increase in phenolics during the extrusion process could be due to the destruction of cell walls, the consequent release of phenolic compounds, and the Maillard reaction products quantified as phenolic compounds [106]. In another study, a significant decrease of total polyphenols and antioxidant activity was observed in a bean/corn mixture during extrusion. However, this decrease was attributed to the process conditions [107]. In a different work, extruded sorghum did not cause the loss of condensed tannins but made them difficult to extract [108]. In general, a reduction of the phenolic content caused by the thermal process resulted from the polymerization of these compounds and consequently less extractability [51]. Several researchers have shown either the retention or increase of bioactive compounds during the extrusion cooking in cereals such as wheat, barley, rye, and oat-based products [97, 109].

Phenolic compounds during extrusion may undergo decarboxylation due to high barrel temperatures. Also, high moisture content may promote polymerization of phenols and tannins reducing their extractability and antioxidant activity [110]. While the increase in phenolic acids in extruded products is generally due to the release from the cell wall matrix, most bioactive compounds are temperature sensitive, and barrel temperature plays a significant role in the stability of their antioxidant properties [6]. In a study with pigmented (white, yellow, blue, and red) Mexican maize processed by extrusion cooking, the flours obtained were used to make tortillas. These tortillas showed high retention of total phenolic content (76–93%), total ferulic acid (58–97%), and antioxidant activity evaluated as ORAC assay (93–75%) compared to tortillas made with the traditional process [29]. These results clearly indicate that extrusion cooking is an alternative to obtain products with higher levels of phytochemicals and antioxidant activity. Similarly, Aguayo-Rojas et al. [32] found higher retention in total phenolic content (76–87%) and antioxidant activity (ORAC, 87–90%), in tortillas from pigmented Mexican maize elaborated from extruded flours.

During extrusion cooking, a significant decrease (<50%) in total anthocyanins in blue Mexican maize has been reported. This decrement is mainly attributed to the flavonoids' sensitivity to high temperatures [29, 32, 111]. However, there is also an increase in biologically important monomers and dimers due to the disruption of the cell wall food matrix [112].

3.3.2. Other bioactive compounds

Carotenoids and isoflavones are also affected by extrusion cooking. In a report using a corn/soy blend, the extrusion barrel temperature and the moisture content showed an increase in the acetyl derivatives of genistein and daidzein and a decrease in malonyl analogues indicating thermal decarboxylation [113]. In a study with eight genotypes of creole Mexican maize (yellow and red) processed into extruded flours and tortillas, the total carotenoid content showed a retention range of 69–79% with respect to raw maize. Likewise, the concentration of the individual carotenoid compounds (Lutein, Zeaxanthin, β-cryptoxanthin, and β-carotene) decreased with the extrusion process. Interestingly, lutein, the major carotenoid in maize, showed an average retention of 60–71% with respect to raw maize. The significant loss on levels and profiles of carotenoids and lipophilic antioxidant activity during the elaboration or tortillas could be attributed mainly to the effect of thermal process, which induce carotenoid degradation and reactions such as isomerization and oxidation [114]. In light of the above, it is prudent to conclude that the effect of extrusion on bioactive compounds is not only dependent of the grain variety, but it is also important to select the appropriate processing conditions.

4. Future perspectives and conclusions

The consumption of whole grains is considered to have significant health benefits in prevention from chronic diseases such as cardiovascular disease, diabetes, and cancer. Cereal grains undergo physical and chemical changes during processing, so careful considerations should be taken to minimize or prevent any unfavorable changes in nutritional properties and the content of bioactive compounds. Most of the bioactive and phenolic compounds are mainly concentrated in the outer layer of cereal grains, and thus consuming whole-grain products is considered the best solution to increase the health benefits of cereal products. Extruders can be used to cook, form, mix, texturize, and shape food products under conditions that favor quality retention, high productivity, and low cost. In this regard, extrusion cooking is an ideal method for manufacturing a number of cereal products and to produce whole-grain products maintaining all the anatomic parts of the grain.

Despite the importance of selecting the appropriate food-processing conditions to improve the nutritional characteristics and increase the amount of biocomponents of the final product; research in this area is still limited. Thus, novel studies focusing on the optimization of thermal and nonthermal operations during extrusion have a vast potential for the food industry. Processing operations optimized for food safety may be combined with phytochemical studies to analyze both the nutritional and safety aspects. Finally, extrusion cooking has a potential for becoming the most important food-processing technology in the future, which can potentially be exploited.

Author details

Cuauhtémoc Reyes Moreno, Perla C. Reyes Fernández, Edith O. Cuevas Rodríguez, Jorge Milán Carrillo and Saraid Mora Rochín*

*Address all correspondence to: smora@uas.edu.mx

Faculty of Chemical and Biological Sciences, Autonomous University of Sinaloa, Mexico

References

[1] Singh S, Gamlath S, Wakeling L. Nutritional aspects of food extrusion: A review. International Journal of Food Science & Technology. 2007;**42**(8):916-929

[2] Nayak B, Liu RH, Tang J. Effect of processing on phenolic antioxidants of fruits, vegetables, and grains—A review. Critical reviews in Food Science and Nutrition. 2015;**55**(7): 887-918

[3] Riaz MN, Asif M, Ali R. Stability of vitamins during extrusion. Critical Reviews in Food Science and Nutrition. 2009;**49**(4):361-368

[4] Guy R. Extrusion Cooking: Technologies and Applications. 1st ed. Woodhead Publishing Ltd, Cambridge; 2001. 288 p. ISBN: 9781855736313

[5] Gbenyi D, Nkama I, Badau M, Idakwo P. Effect of extrusion conditions on nutrient status of ready-to-eat breakfast cereals from sorghum-cowpea extrudates. Journal Food Processing & Beverages. 2016;**4**(2):8

[6] Brennan C, Brennan M, Derbyshire E, Tiwari BK. Effects of extrusion on the polyphenols, vitamins and antioxidant activity of foods. Trends in Food Science & Technology. 2011;**22**(10):570-575

[7] Sarwar MH, Sarwar MF, Sarwar M, Qadri NA, Moghal S. The importance of cereals (*Poaceae*: Gramineae) nutrition in human health: A review. Journal of Cereals and Oilseeds. 2013;**4**(3):32-35

[8] Koehler P, Wieser H. Chemistry of Cereal Grains. In: Gobetti M, Gaenzle M, editors. Handbook of Sourdough Biotechnology. 1st ed. Springer, New York; 2013. p. 11-45. DOI: 10.1007/978-1-4614-5425-0_2

[9] Allowances RD. Food and Nutrition Board. National Research Council. 10th ed., Washington, DC: National Academy of Sciences; 1989

[10] Reyes-Moreno C, Ayala-Rodríguez AE, Milán-Carrillo J, Mora-Rochín S, López-Valenzuela JA, Valdez-Ortiz A, Paredes-López O, Gutiérrez-Dorado R. Production of nixtamalized flour and tortillas from amarantin transgenic maize lime-cooked in a thermoplastic extruder. Journal of Cereal Science. 2013;**58**(3):465-471

[11] Wang T, He F, Chen G. Improving bioaccessibility and bioavailability of phenolic compounds in cereal grains through processing technologies: A concise review. Journal of Functional Foods. 2014;**7**:101-111

[12] Fardet A. New hypotheses for the health-protective mechanisms of whole-grain cereals: What is beyond fibre? Nutrition Research Reviews. 2010;**23**(01):65-134

[13] Liu RH. Whole grain phytochemicals and health. Journal of Cereal Science. 2007;**46**(3): 207-219

[14] Okarter N, Liu RH. Health benefits of whole grain phytochemicals. Critical Reviews in Food Science and Nutrition. 2010;**50**(3):193-208

[15] Khush G. Productivity improvements in rice. Nutritrion Reviews. 2003;**61**(6 Pt 2): S114–S116

[16] Singhal P, Kaushik G. Therapeutic effect of cereal grains: A review. Critical Reviews in Food Science and Nutrition. 2016;**56**(5):748-759

[17] Slavin JL, Jacobs D, Marquart L, Wiemer K. The role of whole grains in disease prevention. Journal of the American Dietetic Association. 2001;**101**(7):780-785

[18] Wójtowicz A, Mitrus M, Oniszczuk T, Mościcki L, Kręcisz M, Oniszczuk A. Selected physical properties, texture and sensory characteristics of extruded breakfast cereals based on wholegrain wheat flour. Agriculture and Agricultural Science Procedia. 2015;**7**:301-308

[19] Adom KK, Sorrells ME, Liu RH. Phytochemicals and antioxidant activity of milled fractions of different wheat varieties. Journal of Agricultural and Food Chemistry. 2005;**53**(6):2297-2306

[20] Eliasson AC, Larsson K. Basic concepts of surface and colloid chemistry, In: Ealiasson AC, Larson K, editors. Cereals in Breadmaking: A Molecular Colloidal Approach. New York: Marcel Dekker; 1993. 1 p

[21] Carlsen MH, Halvorsen BL, Holte K, Bøhn SK, Dragland S, Sampson L, Willey C, Senoo H, Umezono Y, Sanada C, Barikmo I, Berhe N, Willet WC, Phillips KM, Jacobs DR, Blomhoff R. The total antioxidant content of more than 3100 foods, beverages, spices, herbs and supplements used worldwide. Nutrition Journal. 2010;**9**(1):3

[22] Dasgupta A, Klein K. Antioxidants in Food, Vitamins and Supplements: Prevention and Treatment of Disease. 1st ed. Elsevier, USA; 2014. p. 316. DOI: 10.1016/B978-0-12-405 872-9.00019-7

[23] Sarkar D, Shetty K. Metabolic stimulation of plant phenolics for food preservation and health. Annual Review of Food Science and Technology. 2014;**5**:395-413

[24] Van Hung P. Phenolic compounds of cereals and their antioxidant capacity. Critical Reviews in Food Science and Nutrition. 2016;**56**(1):25-35

[25] Ragaee S, Seetharaman K, Abdel-Aal E-SM. The impact of milling and thermal processing on phenolic compounds in cereal grains. Critical Reviews in Food Science and Nutrition. 2014;**54**(7):837-849

[26] Liu RH. Dietary bioactive compounds and their health implications. Journal of Food Science. 2013;**78**(s1):A18-A25

[27] Adom KK, Liu RH. Antioxidant activity of grains. Journal of Agricultural and Food Chemistry. 2002;**50**(21):6182-6187

[28] Andreasen MF, Christensen LP, Meyer AS, Hansen Å. Content of phenolic acids and ferulic acid dehydrodimers in 17 Rye (*Secale cereale* L.) Varieties. Journal of Agricultural and Food Chemistry. 2000;**48**(7):2837-2842

[29] Mora-Rochin S, Gutiérrez-Uribe JA, Serna-Saldivar SO, Sánchez-Peña P, Reyes-Moreno C, Milán-Carrillo J. Phenolic content and antioxidant activity of tortillas produced from pigmented maize processed by conventional nixtamalization or extrusion cooking. Journal of Cereal Science. 2010;**52**(3):502-508

[30] Chandrasekara A, Shahidi F. Determination of antioxidant activity in free and hydrolyzed fractions of millet grains and characterization of their phenolic profiles by HPLC-DAD-ESI-MS. Journal of Functional Foods. 2011;**3**(3):144-158

[31] Faraj A, Vasanthan T, Hoover R. The effect of extrusion cooking on resistant starch formation in waxy and regular barley flours. Food Research International. 2004;**37**(5):517-525

[32] Aguayo-Rojas J, Mora-Rochín S, Cuevas-Rodríguez EO, Serna-Saldivar SO, Gutierrez-Uribe JA, Reyes-Moreno C, Milán-Carrillo J. Phytochemicals and antioxidant capacity of tortillas obtained after lime-cooking extrusion process of whole pigmented mexican maize. Plant Foods for Human Nutrition. 2012;**67**(2):178-185

[33] Alam M, Kaur J, Khaira H, Gupta K. Extrusion and extruded products: Changes in quality attributes as affected by extrusion process parameters: A review. Critical Reviews in Food Science and Nutrition. 2016;**56**(3):445-473

[34] Milán-Carrillo J, Gutiérrez-Dorado R, Perales-Sánchez JXK, Cuevas-Rodríguez EO, Ramírez-Wong B, Reyes-Moreno C. The optimization of the extrusion process when using maize flour with a modified amino acid profile for making tortillas. International Journal of Food Science & Technology. 2006;**41**(7):727-736

[35] Chang YH, Ng PK. Effects of Extrusion process variables on extractable ginsenosides in wheat– ginseng extrudates. Journal of Agricultural and Food Chemistry. 2009;**57**(6):2356-2362

[36] Hauck B, Huber G. Single Screw vs Twin Screw Extrusion. Cereal Foods World (USA); 1989

[37] Castells M, Marin S, Sanchis V, Ramos A. Fate of mycotoxins in cereals during extrusion cooking: A review. Food Additives and Contaminants. 2005;**22**(2):150-157

[38] De Pilli T, Severini C, Carbone BF, Giuliani R, Derossi A. Improving fatty extrudate structure with amylase and protease. Journal of Food Biochemistry. 2004;**28**(5):387-403

[39] Meuser CayL, B. System analytical model for the extrusion of starches. Food Extrusion Science and Technology. New York: Marcel Dekker Inc.; 1992

[40] Harper J. Food extruders and their applications. In: Mercier C, Linko P. Extrusion Cooking. St. Paul, MN: Am Assoc Cereal Chemistry Press; 1989

[41] Eastman J, Orthoefer F, Solorio S. Using extrusion to create breakfast cereal products. Cereal Foods World. 2001;**46**(10):468-471

[42] Leyva-Corral J, Quintero-Ramos A, Camacho-Dávila A, de Jesús Zazueta-Morales J, Aguilar-Palazuelos E, Ruiz-Gutiérrez MG, Meléndrez-Pizarro CO. Polyphenolic compound stability

and antioxidant capacity of apple pomace in an extruded cereal. LWT-Food Science and Technology. 2016;**65**:228-236

[43] Bookwalter G, Mustakas G, Kwolek W, McGhee J, Albrecht W. Full-fat soy flour extrusion cooked: Properties and food uses. Journal of Food Science. 1971;**36**(1):5-9

[44] Harper JM. Extrusion of Foods, Vol. I, CRC Press, Florida; 1981. p. 212. ISBN 0-8493-5204-5

[45] Bhattacharya S, Prakash M. Extrusion of blends of rice and chick pea flours: A response surface analysis. Journal of Food Engineering. 1994;**21**(3):315-330

[46] Riedl KM, Hagerman AE. Tannin– protein complexes as radical scavengers and radical sinks. Journal of Agricultural and Food Chemistry. 2001;**49**(10):4917-4923

[47] Žilić S, Mogol BA, Akıllıoğlu G, Serpen A, Delić N, Gökmen V. Effects of extrusion, infrared and microwave processing on Maillard reaction products and phenolic compounds in soybean. Journal of the Science of Food and Agriculture. 2014;**94**(1):45-51

[48] Grela ER, Jensen SK, Jakobsen K. Fatty acid composition and content of tocopherols and carotenoids in raw and extruded grass pea (*Lathyrus sativus* L). Journal of the Science of Food and Agriculture. 1999;**79**(15):2075-2078

[49] Chiu HW, Peng JC, Tsai SJ, Lui W-B. Effect of extrusion processing on antioxidant activities of corn extrudates fortified with various Chinese yams (*Dioscorea* sp.). Food and Bioprocess Technology. 2012;**5**(6):2462-2473

[50] Remy S, Fulcrand H, Labarbe B, Cheynier V, Moutounet M. First confirmation in red wine of products resulting from direct anthocyanin–tannin reactions. Journal of the Science of Food and Agriculture. 2000;**80**(6):745-751

[51] Dlamini NR, Taylor JR, Rooney LW. The effect of sorghum type and processing on the antioxidant properties of African sorghum-based foods. Food Chemistry. 2007;**105**(4): 1412-1419

[52] Dewanto V, Wu X, Liu RH. Processed sweet corn has higher antioxidant activity. Journal of Agricultural and Food Chemistry. 2002;**50**(17):4959-4964

[53] Awika JM, Dykes L, Gu L, Rooney LW, Prior RL. Processing of sorghum (Sorghum bicolor) and sorghum products alters procyanidin oligomer and polymer distribution and content. Journal of Agricultural and Food Chemistry. 2003;**51**(18):5516-5521

[54] Reyes LF, Villarreal JE, Cisneros-Zevallos L. The increase in antioxidant capacity after wounding depends on the type of fruit or vegetable tissue. Food Chemistry. 2007;**101**(3): 1254-1262

[55] Zieliński H, Michalska A, Piskuła MK, Kozłowska H. Antioxidants in thermally treated buckwheat groats. Molecular Nutrition & Food Research. 2006;**50**(9):824-832

[56] Brennan MA, Derbyshire E, Tiwari BK, Brennan CS. Ready-to-eat snack products: The role of extrusion technology in developing consumer acceptable and nutritious snacks. International Journal of Food Science & Technology. 2013;**48**(5):893-902

[57] Camire M. Extrusion cooking. In: Henry CJK, Chapman C, editors. The Nutrition Handbook for Food Processors. 1st ed. Woodhead Publishing Ltd, Cambridge; 2002. p. 314-326. DOI: 10.1016/B978-1-85573-464-7.50002-9

[58] Viscidi KA, Dougherty MP, Briggs J, Camire ME. Complex phenolic compounds reduce lipid oxidation in extruded oat cereals. LWT-Food Science and Technology. 2004;**37**(7): 789-796

[59] Thachil MT, Chouksey MK, Gudipati V. Amylose-lipid complex formation during extrusion cooking: Effect of added lipid type and amylose level on corn-based puffed snacks. International Journal of Food Science & Technology. 2014;**49**(2):309-316

[60] Camire ME, Dougherty MP, Briggs JL. Antioxidant-rich foods retard lipid oxidation in extruded corn. Cereal Chemistry. 2005;**82**(6):666-670

[61] Sproston MJ, Akoh CC. Antioxidative Effects of a glucose-cysteine maillard reaction product on the oxidative stability of a structured lipid in a complex food emulsion. Journal of Food Science. 2016;**81**(12):2923-2931

[62] Avery SV. Molecular targets of oxidative stress. Biochemical Journal. 2011;**434**(2):201-210

[63] Suzuki H, Chung BS, Isobe S, Hayakawa S, Wada S. Changes in ω-3 polyunsaturated fatty acids in the chum salmon muscle during spawning migration and extrusion cooking. Journal of Food Science. 1988;**53**(6):1659-1661

[64] Ramos Diaz JM, Sundarrajan L, Kariluoto S, Lampi A-M, Tenitz S, Jouppila K. Effect of extrusion cooking on physical properties and chemical composition of corn-based snacks containing amaranth and quinoa: Application of partial least squares regression. Journal of Food Process Engineering. 2017;**40**(1):1-15.

[65] Bhatnagar S, Hanna MA. Amylose-lipid complex formation during single-screw extrusion of various corn starches. Cereal Chemistry. 1994;**71**(6):582-586

[66] Day L. Proteins from land plants–potential resources for human nutrition and food security. Trends in Food Science & Technology. 2013;**32**(1):25-42

[67] Consultation R. Dietary protein quality evaluation in human nutrition. FAO Food and Nutrition Paper; 2011;**92**:3-5.

[68] Duodu K, Taylor J, Belton P, Hamaker B. Factors affecting sorghum protein digestibility. Journal of Cereal Science. 2003;**38**(2):117-131

[69] Sun-Waterhouse D, Zhao M, Waterhouse GI. Protein modification during ingredient preparation and food processing: Approaches to improve food processability and nutrition. Food and Bioprocess Technology. 2014;**7**(7):1853-1893

[70] Vaz L, Arêas JAG. Recovery and upgrading bovine rumen protein by extrusion: Effect of lipid content on protein disulphide cross-linking, solubility and molecular weight. Meat Science. 2010;**84**(1):39-45

[71] Verhoeckx KC, Vissers YM, Baumert JL, Faludi R, Feys M, Flanagan S, Herouet-Guicheney C, Holzhauser T, Shimojo R, van der Bolt N, Wichers H, Kimber I. Food processing and allergenicity. Food and Chemical Toxicology. 2015;**80**:223-240

[72] Camire ME. Chemical changes during extrusion cooking. In: Shahidi F, Ho CT, and van Chuyen N editors. Process-Induced Chemical Changes in Food. Advances in Experimental Medicine and Biology: Springer; 1998; **434**:109-121. DOI: 10.1007/978-1-4899-1925-0_11

[73] Ott C, Jacobs K, Haucke E, Navarrete Santos A, Grune T, Simm A. Role of advanced glycation end products in cellular signaling. Redox Biology. 2014;**2**:411-429

[74] Kellow NJ, Coughlan MT. Effect of diet-derived advanced glycation end products on inflammation. Nutrition Reviews. 2015;**73**(11):737-759

[75] Poulsen MW, Hedegaard RV, Andersen JM, de Courten B, Bugel S, Nielsen J, Skibsted LH, Dragsted LO. Advanced glycation endproducts in food and their effects on health. Food and Chemical Toxicology. 2013;**60**:10-37

[76] Ogasawara M, Katsumata T, Egi M. Taste properties of Maillard-reaction products prepared from 1000 to 5000Da peptide. Food Chemistry. 2006;**99**(3):600-604

[77] Hurrell R. Influence of the Maillard reaction on the nutritional value of foods. In: Finot PA, Aeschabacher HU, Hurrell R, Liardon R, editors. The Maillard Reaction in Food Processing, Human Nutrition and Physiology. 1st ed. Birkhäuser Verlag, 1990. p. 245-258. ISBN: 9783764323547

[78] Butterworth PJ, Warren FJ, Ellis PR. Human α-amylase and starch digestion: An interesting marriage. Starch-Stärke. 2011;**63**(7):395-405

[79] Lai LS, Kokini JL. Physicochemical changes and rheological properties of starch during extrusion. A review. Biotechnology Progress. 1991;**7**(3):251-266

[80] Brennan MA, Merts I, Monro J, Woolnough J, Brennan CS. Impact of guar and wheat bran on the physical and nutritional quality of extruded breakfast cereals. Starch-Stärke. 2008;**60**(5):248-256

[81] Borejszo ZB, Khan KH. Reduction of flatulence-causing sugars by high temperature extrusion of pinto bean high starch fractions. Journal of Food Science. 1992;**57**(3):771-777

[82] Alonso R, Orue E, Zabalza MJ, Grant G, Marzo F. Effect of extrusion cooking on structure and functional properties of pea and kidney bean proteins. Journal of the Science of Food and Agriculture. 2000;**80**(3):397-403

[83] Mahasukhonthachat K, Sopade P, Gidley M. Kinetics of starch digestion and functional properties of twin-screw extruded sorghum. Journal of Cereal Science. 2010;**51**(3):392-401

[84] Haralampu S. Resistant starch—a review of the physical properties and biological impact of RS 3. Carbohydrate Polymers. 2000;**41**(3):285-292

[85] Blaak E, Antoine JM, Benton D, Björck I, Bozzetto L, Brouns F, Diamant M, Dye L, Hulshor T, Holst JJ, Lamport DJ, Laville M,Lawton CL, Meheust A, Nilson A, Normand S, Rivellsese AA, Theis S, Torekov SS, Vinoy S. Impact of postprandial glycaemia on health and prevention of disease. Obesity Reviews. 2012;**13**(10):923-984

[86] Alsaffar AA. Effect of food processing on the resistant starch content of cereals and cereal products–a review. International Journal of Food Science & Technology. 2011;**46**(3):455-462

[87] Chanvrier H, Uthayakumaran S, Appelqvist IA, Gidley MJ, Gilbert EP, López-Rubio A. Influence of storage conditions on the structure, thermal behavior, and formation of enzyme-resistant starch in extruded starches. Journal of Agricultural and Food Chemistry. 2007;**55**(24):9883-9890

[88] Huth M, Dongowski G, Gebhardt E, Flamme W. Functional properties of dietary fibre enriched extrudates from barley. Journal of Cereal Science. 2000;**32**(2):115-128

[89] Jing Y, Chi Y-J. Effects of twin-screw extrusion on soluble dietary fibre and physicochemical properties of soybean residue. Food Chemistry. 2013;**138**(2):884-889

[90] Chen Y, Ye R, Yin L, Zhang N. Novel blasting extrusion processing improved the physicochemical properties of soluble dietary fiber from soybean residue and in vivo evaluation. Journal of Food Engineering. 2014;**120**:1-8

[91] Stojceska V, Ainsworth P, Plunkett A, İbanoğlu Ş. The advantage of using extrusion processing for increasing dietary fibre level in gluten-free products. Food Chemistry. 2010;**121**(1):156-164

[92] Vasanthan T, Gaosong J, Yeung J, Li J. Dietary fiber profile of barley flour as affected by extrusion cooking. Food Chemistry. 2002;**77**(1):35-40

[93] Esposito F, Arlotti G, Maria Bonifati A, Napolitano A, Vitale D, Fogliano V. Antioxidant activity and dietary fibre in durum wheat bran by-products. Food Research International. 2005;**38**(10):1167-1173

[94] Athar N, Hardacre A, Taylor G, Clark S, Harding R, McLaughlin J. Vitamin retention in extruded food products. Journal of Food Composition and Analysis. 2006;**19**(4):379-383

[95] Boyaci BB, Han J-Y, Masatcioglu MT, Yalcin E, Celik S, Ryu G-H, et al. Effects of cold extrusion process on thiamine and riboflavin contents of fortified corn extrudates. Food Chemistry. 2012;**132**(4):2165-2170

[96] Tiwari U, Cummins E. Nutritional importance and effect of processing on tocols in cereals. Trends in Food Science & Technology. 2009;**20**(11):511-520

[97] Zielinski H, Kozlowska H, Lewczuk B. Bioactive compounds in the cereal grains before and after hydrothermal processing. Innovative Food Science & Emerging Technologies. 2001;**2**(3):159-169

[98] Camire ME, Camire A, Krumhar K. Chemical and nutritional changes in foods during extrusion. Critical Reviews in Food Science & Nutrition. 1990;**29**(1):35-57

[99] Alonso R, Rubio L, Muzquiz M, Marzo F. The effect of extrusion cooking on mineral bioavailability in pea and kidney bean seed meals. Animal Feed Science and Technology. 2001;**94**(1):1-13

[100] Andersson Y, Hedlund B, Jonsson L, Svensson S. Extrusion cooking of a high-fiber cereal product with crispbread character [Wheat bran, secondary starch, and gluten]. Cereal Chemistry (USA). 1981;**58**(5):370-374

[101] Nicoli M, Anese M, Parpinel M. Influence of processing on the antioxidant properties of fruit and vegetables. Trends in Food Science & Technology. 1999;**10**(3):94-100

[102] Altan A, McCarthy KL, Maskan M. Effect of extrusion process on antioxidant activity, total phenolics and β-glucan content of extrudates developed from barley-fruit and vegetable by-products. International Journal of Food Science & Technology. 2009;**44**(6):1263-1271

[103] Liu RH. Potential synergy of phytochemicals in cancer prevention: Mechanism of action. The Journal of Nutrition. 2004;**134**(12):3479-3485

[104] Dimberg L, Molteberg E, Solheim R, Frølich W. Variation in oat groats due to variety, storage and heat treatment. I: Phenolic compounds. Journal of Cereal Science. 1996;**24**(3):263-272

[105] Şensoy Í, Rosen RT, Ho C-T, Karwe MV. Effect of processing on buckwheat phenolics and antioxidant activity. Food Chemistry. 2006;**99**(2):388-393

[106] Espinoza-Moreno RJ, Reyes-Moreno C, Milán-Carrillo J, López-Valenzuela JA, Paredes-López O, Gutiérrez-Dorado R. Healthy ready-to-eat expanded snack with high nutritional and antioxidant value produced from whole amarantin transgenic maize and black common bean. Plant Foods for Human Nutrition. 2016;**71**(2):218-224

[107] Delgado-Licon E, Ayala ALM, Rocha-Guzman NE, Gallegos-Infante J-A, Atienzo-Lazos M, Drzewiecki J, Martínez-Sánchez CE, Gorinstein S. Influence of extrusion on the bioactive compounds and the antioxidant capacity of the bean/corn mixtures. International Journal of Food Sciences and Nutrition. 2009;**60**(6):522-532

[108] Awika JM, Rooney LW, Wu X, Prior RL, Cisneros-Zevallos L. Screening methods to measure antioxidant activity of sorghum (*Sorghum bicolor*) and sorghum products. Journal of Agricultural and Food Chemistry. 2003;**51**(23):6657-6662

[109] Baublis A, Clydesdale F, Decker E. Antioxidants in wheat-based breakfast cereals. Cereal Foods World. 2000;**45**(2):71-74

[110] Repo-Carrasco-Valencia R, de La Cruz AA, Alvarez JCI, Kallio H. Chemical and functional characterization of kaniwa (*Chenopodium pallidicaule*) grain, extrudate and bran. Plant Foods for Human Nutrition. 2009;**64**(2):94-101

[111] Sánchez-Madrigal MÁ, Quintero-Ramos A, Martínez-Bustos F, Meléndez-Pizarro CO, Ruiz-Gutiérrez MG, Camacho-Dávila A, Torres-Chávez P, Ramírez-Wong B. Effect of different calcium sources on the bioactive compounds stability of extruded and nixtamalized blue maize flours. Journal of Food Science and Technology. 2015;**52**(5):2701-2710

[112] Khanal R, Howard L, Prior R. Procyanidin content of grape seed and pomace, and total anthocyanin content of grape pomace as affected by extrusion processing. Journal of Food Science. 2009;**74**(6):H174–H182

[113] Mahungu S, Diaz-Mercado S, Li J, Schwenk M, Singletary K, Faller J. Stability of isoflavones during extrusion processing of corn/soy mixture. Journal of Agricultural and Food Chemistry. 1999;**47**(1):279-284

5

Extrusion from Gear Manufacturing to Production of Cereals

Sayyad Zahid Qamar

1. Introduction

Extrusion is a manufacturing process that generates profiles having a fixed cross-sectional area by pushing the billet material through a die. It is a bulk-forming process, where the stock material is deformed into the final shape under pressure (and temperature), without any melting and solidification or material removal. Its various major advantages over other manufacturing processes include huge shape versatility (from simple to extremely complex profiles); ability to handle brittle materials, as only compressive and shear stresses are employed during deformation; and very good surface finish of the extrudate (almost no postprocessing finishing operations required). By adding a mandrel to the die set, complex hollow shapes can also be extruded. The process is suitable for both hot and cold working. Metals and alloys, polymers and plastics, ceramics, concrete, and food products are the materials that are generally extruded. Extrusion of aluminum alloys finds extensive applications in construction, automobile, aerospace, and other sectors [1–3, 5].

2. Metal extrusion

The first patent for extrusion (called "squirting" at that time) was filed in 1797 by Joseph Bramah for the production of pipes from metals of low hardness. Preheated metal was forced through a die using a hand-worked plunger. Thomas Burr modified the process in 1820 to produce lead pipes, using a hydraulic press (which was also invented by Bramah). The method was expanded in 1894 by Alexander Dick to manufacture products made of copper and brass alloys [4].

Metals that are commonly extruded include aluminum, brass, copper, lead and tin, magnesium, zinc, steel, and titanium. In the case of steels, extrusion is generally restricted to plain-carbon steels; alloy steels and stainless steels are not suitable for this process. Some of the common extruded products are tracks, frames, rails, rods, bars, pipes and pipe fittings, wire and cable, tubes, hardware components and fittings, automobile parts, aircraft components, parts for nuclear industry, and a variety of general engineering and structural parts [4].

In commercial extrusion, billet and die set are preheated and then loaded into the container, which is also held at a suitable temperature. A dummy block is placed behind the billet, and the ram pressure forces the material through the die set. The whole equipment is called the extrusion press. After extrusion, the product is stretched (to straighten it out) and cooled (to reduce the warm softness), cut into required length, and stacked. Heat treatment (age hardening), anodizing and/or painting are done if required [5, 6]. Major components of the extrusion press are shown in **Figure 1**, while the sequence of operations in a typical commercial extrusion plant is shown schematically in **Figure 2**.

2.1. Extrusion of spur gears

Manufacturing of gears depends on the machinery available, customer requirements (design specifications), cost of production, and type of material from which the gear is to be made. There are many methods for manufacturing gears, including metal removal (machining), casting, tamping and fine blanking, cold drawing and extrusion, powder metallurgy, injection molding, gear rolling, and forging. Most of these processes are suited for gears with low wear requirements, low power transmission, and relatively low accuracy of transmitted motion. When the application involves higher values of one or more of these characteristics, forged or machined gears are used [7, 8].

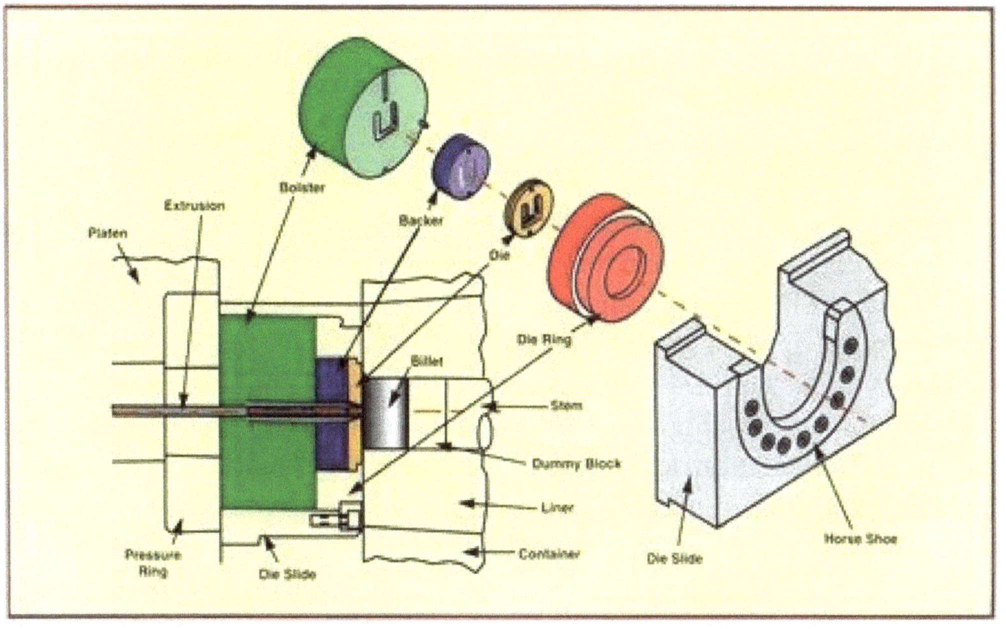

Figure 1. Major components of the metal extrusion press.

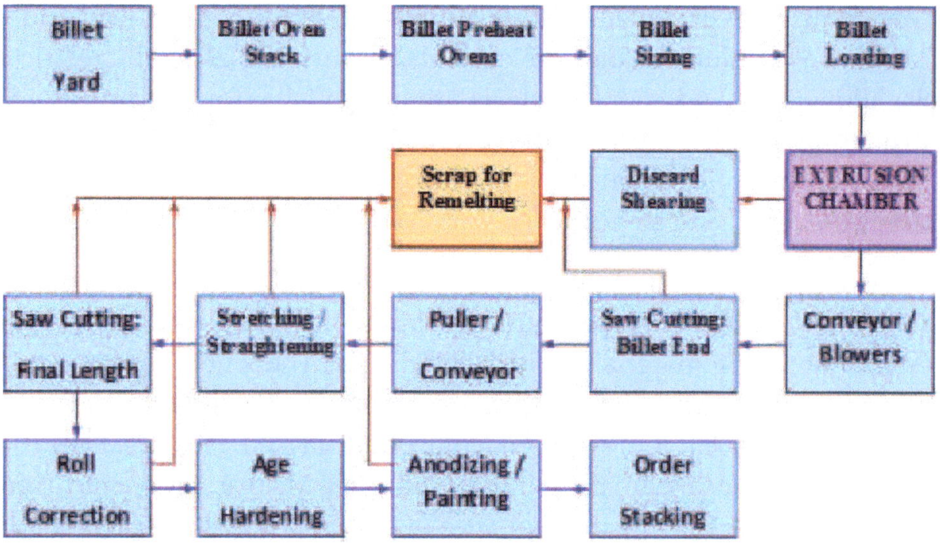

Figure 2. Plant layout (or process flow) in a typical commercial extrusion facility.

The process of cold drawing and extrusion requires the least tool expenditure for mass production of spur gear-toothed gear elements. It is extremely versatile, and almost any desired tooth form can be produced. A bar/billet is pulled (drawn) or pushed (extruded) through a series of several dies, the last having the final shape of the desired tooth form. Since the material is displaced by pressure, the outside surface is work-hardened and quite smooth. After passing through the dies, the bars (or blanks) are known as pinion rods. They are often put into screw machines that finish the individual gears. Any material that has good drawing properties, such as high-carbon steels, brass, bronze, aluminum, and stainless steel, may be used for the drawn pinion rod [9, 10].

Gears and pinions manufactured by this process have a large variety of applications and have been used on watches, electric clocks, spring-wound clocks, typewriters, carburettors, magnetos, small motors, switch apparatus, taximeters, cameras, slot machines, all types of mechanical toys, and many other parts for machinery of all kinds [11, 12].

2.2. Solid tubular expansion

In "direct extrusion," the ram pushes the billet material through the die cavity in such a way that the ram and the product move in the same direction. During "indirect extrusion," the die remains stationary, while the billet and the container move together, and the movement of the ram and the billet is in opposite directions. One novel adaptation of indirect extrusion is the process of tube expansion [13]. In solid expandable tubular (SET) technology, a metal tube (or pipe) is expanded by pushing (or pulling) a conical mandrel (die) through it [14]. Unlike commercial direct extrusion carried at higher temperatures, tube expansion is a type of cold (room temperature) extrusion process. In-situ expansion of petroleum tubulars has recently found many applications in the oil and gas drilling and development industry [15], especially for enhanced oil recovery (EOR). SET-based techniques such as water shutoff, zonal isolation, slim well design, etc. make production possible from old and depleted wells, and result in huge savings in material, time, and cost.

2.3. Stiff vacuum extrusion/extrusion briquettes

Extrusion can be efficiently utilized as a compaction/agglomeration process. One major application is the production of fuel briquettes. Initial attempts at extrusion-based agglomeration of steel were rather unsuccessful. Bethlehem Steel incorporated an agglomeration unit in their blast furnace in 1994. It was based on the stiff vacuum extrusion technology that was earlier developed by JC Steele & Sons and was generally used before for brickmaking. However, the company and its briquette unit closed down soon thereafter. The group consisting of Bizhanov, Kurunov, et al. is credited with serious studies and technology revival in the field, during 2009–2013. Their metallurgical process was verified through actual-scale industrial experiments and later converted into commercial production. They also coined the term "brex" to differentiate the product from traditional briquettes [16].

There is a growing interest now in agglomeration technology based on stiff extrusion. New factories have been built for briquetting in very large steel and ferrous alloy companies. This extrusion-based technology is used for agglomeration of steel mill wastes (blast furnace dust, blast furnace sludge, and lime plant sludge). The low-cost brex is a good substitute for expensive iron ore feedstock in blast furnaces used for pig iron production. Agglomeration of process dust is also done for recovery of nickel ore through the rotary kiln electric furnace process. Other uses of this technology include compaction of fuel briquettes from coal fines, pelletizing of synthetic gypsum to make fertilizer filler, conversion of low energy lignite coal into a higher energy fuel, pelletizing of bauxite to make high-grade alumina feedstocks for refractory products, production of synthetic gypsum rock using cement kiln dust, etc. [17].

3. Polymer extrusion

A very large number of different types of polymers can be extruded into products such as plastic tubes and pipes, rods, rails, fencing, seals, window frames, deck railings, sheets and films, coatings, wire insulation, etc. In 1820, Thomas Hancock invented a rubber "masticator" designed to reclaim processed rubber scraps. In 1836, Edwin Chaffee developed a two-roller machine to mix additives into rubber [18]. Thermoplastic extrusion was carried out for the first time in 1935 by Paul Troester and Ashley Gershoff in Hamburg, Germany. Shortly after, Roberto Colombo developed the first twin screw extruders in Italy [19]. Though thermoplastics are the main product, elastomers and thermosetting plastics can also be produced through extrusion. Cross-linking required for these polymers is formed as the material is heated and melted in the extruder. Further thermoforming of the extruded thermoplastic may be needed in some cases.

Schematic diagram of typical polymer extrusion is shown in **Figure 3**. The basic components of a plastic extruder are screw, drive, barrel, hopper, and die, the most critical part being the helical screw. It helps in transporting, heating and melting, and mixing of the polymer. The starting material can be in the form of chips, pellets, granules, flakes or powders. These are first dried in a top-mounted hopper to remove the moisture and are then gravity-fed into the extruder barrel containing the feed screw. If needed, additives (colorants, ultraviolet inhibitors, etc.) can be mixed into the resin before reaching the feed hopper in liquid or pellet form. Plastic extruders also find

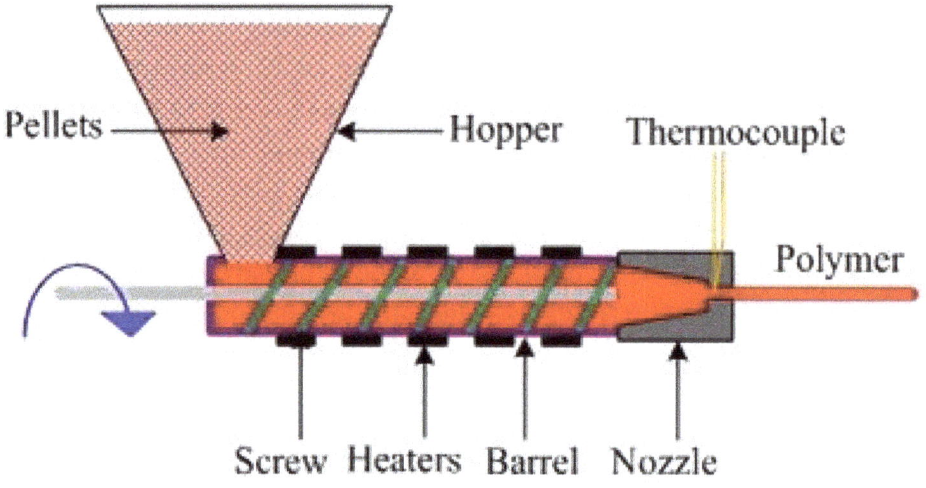

Figure 3. A schematic process diagram of typical polymer extrusion.

extensive use in recycling of plastic waste. This of course needs cleaning and sorting, and blending into the pellet-type resin stock before being used for extrusion [20].

The polymer resin is converted and melted through a combination of heating elements (positioned on the barrel) and shear energy coming from the rotating screw. The melting temperature is maintained with the help of thermocouples. Before entering the die, the resin goes through a reinforced screen (breaker plate) for removal of any contaminants. Screw motion forces the hot resin through the die, generating a continuous polymer product of the desired profile. The puller or "caterpillar haul-off" pulls (or stretches) the extruded material, while also cooling and solidifying it, with the help of blowers or a water bath. This tension is necessary for good quality and can also be applied by a pelletizer, which pulls the strands for later cutting [21].

4. Food extrusion

When extrusion is used for food processing, it is known as food extrusion. Different die shapes are used for different types of food. Extruded food product is cut to the desired length by blades. The extruder consists of a large, rotating screw that fits snugly inside a stationary barrel, fitted with a die set at its outward end [22].

Mass production of food is possible through the continuous and efficient system of extrusion, ensuring a consistent and uniform product. Extrusion has enabled the production of newly processed food products and revolutionized many conventional snack manufacturing processes [23]. Extruded food is usually high in starch content and includes products such as pasta (spaghetti, noodles, macaroni, etc.), breads (croutons, bread sticks and flat breads), breakfast cereals, ready-made snacks, confectionery, etc. [24].

The first food extruder was designed to manufacture sausages in 1870s [25]. Packaged dry pasta and breakfast cereals have been produced via extrusion since 1930s [26]. The method was applied to pet food production in 1950s. It has also been incorporated into kitchen appliances, such as meat grinders, herb grinders, coffee grinders and some types of pasta makers [27]. Extrusion of pet food is shown schematically in **Figure 4**.

4.1. Extrusion cooking

As a food product is processed through extrusion, it is also "cooked." During extrusion cooking, the food product is heated under a high degree of pressure and then slowly forced through a series of pores into another cooking chamber. As this process takes place, the moisture content of the food is reduced significantly, leaving behind a product that is thoroughly cooked and dried. The remaining extruded product is then ready for inclusion in dry mixes or further processing to produce various many of the packaged products that many consumers rely upon today [26]. This wide range of packaged foods includes salty snacks, breakfast bars, cereal-based health foods, etc. Even nongrainy foods such as soybeans are prepared using extrusion cooking, forming the basis for a number of vegetarian products [27].

4.2. Functional food

A functional food is a food that also performs an additional function beyond basic nutrition, generally related to improvement of health or prevention of disease. This is achieved by adding new ingredients or a different proportion of existing ingredients [28]. A familiar example of a functional food is oatmeal because it contains soluble fiber that can help lower the cholesterol levels. An example of a food "modified" to have health benefits is orange juice that has been fortified with calcium for bone health [29].

4.3. Nixtamalization

The process used for the preparation of maize (corn) or similar grain is known as nixtamalization. The grain is soaked and cooked in an alkaline solution (usually limewater), and hulled.

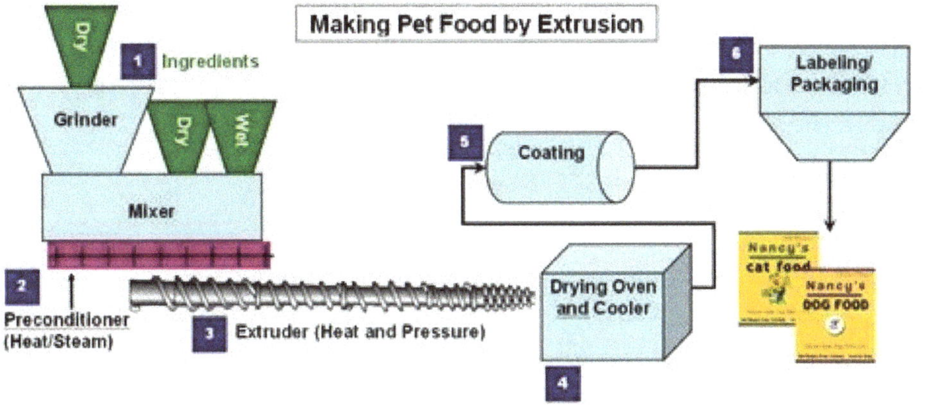

Figure 4. Schematic illustration of the process of pet-food extrusion.

Removal of the pericarp from other grains such as sorghum, using an alkali process, can also be termed as nixtamalization. Maize prepared in this way has several benefits over unprocessed grain: its grinding becomes easier, its nutritional value is increased, its taste and aroma are improved and the amount of mycotoxins is reduced [30].

The high alkalinity of lime and ash helps the dissolution of hemicellulose, the major glue-like component of the maize cell walls, and loosens the hulls from the kernels and softens the maize. Some of the corn oil is broken down into emulsifying agents (monoglycerides and diglycerides), while bonding of the maize proteins to each other is also facilitated. It is difficult to form dough by addition of water using cornmeal made from untreated ground maize. However, the above chemical changes allow easier dough formation. These benefits make nixtamalization a crucial preliminary step for further processing of maize into food products. The process is used together with both traditional and industrial methods used for the production of tortillas and tortilla chips (but not corn chips), tamales, hominy, and many other items [31].

4.4. Aquafeed/fish feed

Production of aquafeed (fish feed) is another application of food extrusion. Aquaculture (rearing/farming of aquatic animals or cultivation of aquatic plants for food) is one of the fastest developing sectors in the world. This growth directly translates into an increasing demand for aquafeed. One major type of aquafeed processing technology is extrusion. The nutrient components in fish feed mainly consist of protein, starch, crude fat, raw fiber, crude ash, vitamin and minerals. Protein provides energy and builds muscles. Vitamins and minerals can enhance natural resistance and feed conversion rate [32].

Extruded aquafeed pellet-making process consists of grinding, mixing, extrusion, drying, coating and cooling. As extrusion is a high-temperature short-time (HTST) heating process, it minimizes the degradation of food nutrients while improving the digestibility of protein and starches. Fish feed extruder is an efficient machine designed for processing floating or sinking aquatic feed just by adjusting the formula. Fish feed extrusion consists of wet extrusion and dry extrusion. The high temperature in dry extrusion is acquired through dissipation of mechanical energy from heated surfaces such as barrel and screw surface, or generated by shear forces between wall and material and screw and material. For wet extrusion, the temperature is achieved through preconditioning and steam injection [33].

Author details

Sayyad Zahid Qamar

Address all correspondence to: sayyad@squ.edu.om

Mechanical and Industrial Engineering Department, Sultan Qaboos University, Al-Khodh, Muscat, Sultanate of Oman

References

[1] Qamar SZ. Modeling and analysis of extrusion pressure and die life for complex aluminum profiles, [PhD thesis]. King Fahd University of Petroleum and Minerals; 2004

[2] Qamar SZ, Sheikh AK, Arif AFM. Modeling and Analysis of Aluminum Extrusion: Process, Tooling, and Defects. LAP Lambert Academic Publishing; 2011

[3] Qamar SZ, Arif AFM, Sheikh AK. Analysis of product defects in a typical aluminum extrusion facility. Materials and Manufacturing Processes. July 2004;**19**(3):391-405

[4] Wikipedia-1. Extrusion. 2017. Available from: https://en.wikipedia.org/wiki/Extrusion [Accessed: Jan 5, 2017]

[5] Qamar SZ, Sheikh AK, Arif AFM, Younas M, Pervez T. Monte Carlo simulation of extrusion die life. Journal of Materials Processing Technology. June 2008;**202**(1-3):96-106

[6] Qamar SZ. FEM study of extrusion complexity and dead metal zone. Archives of Materials Science and Engineering. April 2009;**36**(2):110-117

[7] Gearsolutions. Casting, Forming, and Forging. 2017. Available from: http://www.gearsolutions.com/article/detail/6301/casting-forming-and-forging [Accessed: Feb 20, 2017]

[8] Krenzer TT, Coniglio JW. Gear manufacture, In: ASM Handbook. Vol. 16. Machining, ASM International, 1989, pp. 330-355

[9] Dudley DW. Gear-manufacturing methods. In: Handbook of Practical Gear Design. McGraw-Hill Book Company; 1984. p. 5-86

[10] Cooper C. Alternative gear manufacturing. Gear Technology. July–Aug 1998;**1998**:9-16

[11] Townsend DP, editor. Gears made by dies, In: Dudley's Gear Handbook: The Design, Manufacture, and Application of Gears, 2nd ed., McGraw Hill, 1992, pp. 17.1-17.21

[12] Kuhlmann DJ, Raghupathi PS. Manufacturing of forged and extruded gears. Gear Technology. July–Aug 1990;**1990**:36-45

[13] Pervez T, Seibi AC, Al-Hiddabi SA, Al-Jahwari FK, Qamar SZ, Marketz F. Solid tubular expansion in horizontal wells. SPE paper # 105704. In: 15th SPE Middle East Oil & Gas Show and Conference; Bahrain: Society of Petroleum Engineers; 11-14 March 2007

[14] Qamar SZ, Pervez T, Akhtar M, Al-Kharusi MSM. Design and manufacture of Swell packers: Influence of material behavior. Materials and Manufacturing Processes. 2012;**27**(7):727-732 IF 1.629 Oct 2015

[15] Pervez T, Qamar SZ, Al-Hiddabi SA, Al-Jahwari FK, Marketz F, Al-Houqani S, Velden MVD. Tubular expansion in irregularly shaped boreholes—computer simulation and field measurement. Petroleum Science and Technology. 2011;**29**(7):735-744

[16] Briket-Brex. Extrusion Briquettes (brex). 2017. Available from: http://briket-brex.ru/en/o-kompanii/ [Accessed: Jan 20, 2017]

[17] Dalmia YK, Kurunov IF, Steele RB, Bizhanov AM. Production of a new generation of briquettes and their use in blast-furnace smelting. Metallurgist. 2012;**56**(3-4):164-168

[18] Tadmor Z, Gogos CG. Principles of Polymer Processing. 2nd ed. Wiley; 2006

[19] Wikipedia-2. Plastics Extrusion. 2017. Available from: https://en.wikipedia.org/wiki/Plastics_extrusion [Accessed: Jan 5, 2017]

[20] SubsTech. The Extrusion Process. 2017. Available from: http://www.substech.com [Accessed: Jan 10, 2017]

[21] Rauwendaal C. Polymer Extrusion. 4th revised ed. Hanser; 2001

[22] Wikipedia-3. Food Extrusion. 2017. Available from: https://en.wikipedia.org/wiki/Food_extrusion [Accessed: Jan 2]

[23] Harper JM, Clark P. Food extrusion. CRC Critical Reviews in Food Science and Nutrition. 1979;**11**(2):

[24] Akdogan H. High moisture food extrusion. International Journal of Food Science & Technology. 1999;**34**(3):195-207

[25] Karwe MV. Food extrusion, In: Food Engineering. Vol. 3. Oxford Eolss Publishers; 2008

[26] Wisegeek. Extrusion Cooking. 2017. Available from: http://www.wisegeek.com/what-is-extrusion-cooking.htm [Accessed: Jan 10]

[27] Guy R. Extrusion Cooking: Technologies and Applications. Woodhead Publishing; 2001

[28] Wikipedai-4. Functional Food. 2017. Available from: https://en.wikipedia.org/wiki/Functional_food [Accessed: Jan 15, 2017]

[29] Mayoclinic. Functional Foods. 2017. Available from: http://www.mayoclinic.org/healthy-lifestyle/nutrition-and-healthy-eating/ [Accessed: Jan 1, 2017]

[30] Wikipedia-5. Nixtamalization. 2017. Available from: https://en.wikipedia.org/wiki/Nixtamalization [Accessed: Jan 10]

[31] Sefa-Dedeh S, Cornelius B, Sakyi-Dawson E, Afoakwa EO. Effect of nixtamalization on the chemical and functional properties of maize. Food Chemistry. July 2004;**86**(3):317-324

[32] Fish-Feed-Extruder. Extruded Aquafeed and Pelleted Feed. 2017. Available from: http://fish-feed-extruder.com/Application [Accessed: Jan 1, 2017]

[33] Fishfeedmachinery. Fish Feed Extrusion. 2017. Available from: www.fishfeedmachinery.com/ [Accessed: Jan 20, 2017]

6

The Extrusion Cooking Process for the Development of Functional Foods

Martha G. Ruiz-Gutiérrez,
Miguel Á. Sánchez-Madrigal and
Armando Quintero-Ramos

Abstract

The extrusion cooking technology is applied to the development of instant functional foods. It has advantages of low cost, sustainability, and versatility for production of a wide variety of food products. For formulation of functional foods, bioactive compounds are added to base mixtures, the main sources being fruits, vegetables, cereals, oleaginous plants, legumes, and industrial food by-product such as pomace. These sources provide phenolic compounds such as anthocyanins, flavonols, and procyanidins besides betalains, carotenoids, vitamins, amino acids, and complex polysaccharides such as dietary fiber sources. During the extrusion cooking process, ingredients are mixed, conditioned, and transformed to a melt fluid, thus causing degradation or a release of functional compounds because of structural and chemical changes caused by the effects of some process variables such as temperature, moisture content, screw speed, and inherent factors such as geometrical configuration of the extruder. Retention of bioactive compounds to obtain extruded functional foods is an important topic. The description of degradation by means of mathematical models has been used to determine the impact of process variables on stability and concentrations of certain compounds in final extruded products. These models have been successfully applied, showing a good fit and adequately describing the variability of these compounds in extrusion cooking systems under specific conditions.

Keywords: extrusion, functional food, bioactive compound, stability, functional properties

1. Introduction

Currently, consumers demand nutritious food that provides health benefits. This situation has led to transformation of the food processing industry to ensure provision of healthy foods to consumers recognized as "functional foods." Functional foods were first introduced in Japan in the 1980s. Over the years, several authors and institutions have defined this term differently, but the most common or simple definition is "a food or processed food that contains ingredients that have added positive health benefits beyond the basic nutritional function." These ingredients or components are related to disease prevention and improvement of quality of life. A wide range of bioactive compounds have beneficial effects on human health, including probiotics and prebiotics, dietary fiber, vitamins, minerals, proteins, as well as secondary plant metabolites, which include phenolic acids, flavonoids, alkaloids, terpenoids, and glucosinolates. Besides, natural colorants such as betalains (betacyanins and betaxanthins) as well as carotenoids and anthocyanins are of interest owing to their pigmenting potential and antioxidant properties. These components are interesting because many of them possess antioxidant activity and have been shown to have anti-inflammatory, antibacterial, antiviral, and cancer-protective activities. These bioactive compounds are contained in a great variety of foods; **Table 1** shows some sources of these components.

Stability of bioactive compounds such as betacyanins, betaxanthins, and anthocyanins, which are used as natural pigments, is affected by certain external factors such as temperature, pH, and concentration as well as the presence of oxygen, enzymes, water activity, light, and metals. Temperature is one of the main factors that affect most of these bioactive compounds. As for betalains, it is reported that their degradation increases with the increasing temperature and heating time [2, 14]. Those authors reported pigment degradation (betalains) in encapsulated red cactus pear powder as a function of temperature (80, 100, 120, and 140°C) during an extrusion process. Betalain content (betacyanins and betaxanthins) was affected by

Sources		Bioactive compounds	Reference
Fruits	Apple, grape, cherry, peach, mangos, blueberry, cranberry, raspberry, bilberry, cactus pear	Polyphenols, anthocyanins, carotenoids, flavonoids, betalains (betacyanins, betaxanthins)	[1–6]
Vegetables	Carrot, tomato, onion, cauliflower, broccoli	Carotenoids, lycopene, polyphenols, glucosinolates, vitamins	[7, 8]
Grains	Pigmented corn, oat, barley, wheat, amaranth, bean, rice	Anthocyanins, polyphenols, flavonoids, soluble fiber, L-lysine	[1, 9–12]
Oleaginous by-products	Defatted soybean paste, pumpkin, and defatted sunflower pasta	Proteins, polyphenols, carotenoids	[13]

Table 1. Food sources of bioactive compounds.

temperature, indicating that more than a half of these components were lost during the extrusion cooking process because of chemical changes such as isomerization, decarboxylation, and cleavage [15]. It has also been reported that betacyanin degradation in a betanin solution follows first-order reaction kinetics [14, 16]. In addition, the extrusion process results in losses in total polyphenol content and in antioxidant activity owing to temperature effects, indicating that high temperatures, >80°C, may decompose or alter their molecular structure, for example, may cause decarboxylation of free phenolic acids or formation of insoluble complexes with food components like proteins [17, 18]. The degradation of polyphenols and natural pigments during the extrusion process has been linked to the decrease in antioxidant activity of these components because of their structural changes, specifically because of their ability to donate hydrogen atoms from hydroxyl groups to free radicals [19]. The loss of other natural pigments such as anthocyanins and carotenes under the influence of heat during the extrusion process has also been reported [9, 20, 21]. The relations among total polyphenolic content, antioxidant activity, and degradation of anthocyanins caused by high temperature after the extrusion cooking process in varieties of blue maize flour were also reported by Sánchez-Madrigal et al. [9].

As already described above, pH is another factor that affects bioactive compounds such as natural colorants. Although betalains alter their charge because of a pH change, they are not as susceptible as anthocyanins. This is because in the aqueous phase, anthocyanins exist as a mixture of four molecular species, and the concentration of these forms varies depending on pH. **Table 2** shows the different forms and colors that anthocyanins acquire at different pH levels.

In acidic media (at low pH values), anthocyanins are more stable than in alkaline solutions, where they become more susceptible to degradation, as verified in a study by Sánchez-Madrigal et al. [9], who evaluated the effects of two types of calcium salts—calcium hydroxide, Ca(OH)2, and calcium lactate, C6H10O6Ca—at different pH levels and concentrations, during an extrusion and nixtamalization process. The main results revealed that both salts resulted in changes in anthocyanins and therefore in their color. Anthocyanin concentration decreased as Ca(OH)2 concentration increased but increased as the C6H10O6Ca concentration increased. This phenomenon can be attributed to the pH changes caused by each calcium salt, affecting the stability of anthocyanins in flour, which are more stable in acidic media than in alkaline media [22, 23]. Besides, it was observed that flour color is closely related to the

pH	Molecular species	Color
1–3	Flavylium cation	Red
4–5	Carbinol or pseudo base	Colorless
7–8	Quinoidal base	Blue-violet
>5 (ring opening)	Chalcone	Colorless or light yellow

Brouillard [22]; He and Giusti [23].

Table 2. Molecular species of anthocyanins at different pH levels.

structural changes in anthocyanins because of prevailing pH in the cooking medium, resulting in different colors and hues at different pH levels [22–24]. Similar anthocyanin changes were reported in another study [25] during production of tortilla chips with extruded and nixtamalized flours.

Although during the extrusion cooking process a loss of bioactive compounds prevails, it is also reported that under certain process conditions, phenolic compounds are transformed into more easily extractable forms associated with structural changes occurring in the materials subjected to extrusion, thus increasing the release of bioactive compounds present in the cell wall matrix [26, 27]. In a study by Leyva-Corral et al. [1], during production of an instant extruded cereal with apple pomace, it was found that amounts of certain individual phenolic compounds increased under certain extrusion conditions (temperature and moisture content). For example, p-coumaric acid content increased as the temperature increased at all the moisture levels tested, decreasing at temperatures higher than 144°C. For ferulic acid, they found that an increase in moisture content at any temperature studied increased this compound's concentration in the extrudates, reaching the highest value at 195°C. Meanwhile, rutin and phloridzin contents increased until 140°C at intermediate moisture levels (26–29%) and then decreased at higher temperatures.

Knowing the factors affecting the different components that form a food matrix, such as composition, intrinsic characteristics (pH and ionic strength, among others) combined with some extrusion process variables may help a technologist to influence the characteristics or final properties of the desired product. In addition, the presence, incorporation, and stability of bioactive compounds for the development of functional extruded products are topics that must be studied to meet consumer demands.

2. Importance of extrusion processes in product development

The processing methods designed to produce functional foods are diverse and technologically different because they depend on the type of product to be developed. Consequently, there are different kinds of functional food products such as beverages and semisolid or solid foods. These can be obtained by processing methods such as thermal processes, drying technology, freezing processes, and minimal processing technologies. Each one has advantages and technological limitations (in terms of the development of food products), which have an impact on the cost and consumer preferences. An alternative in food processing is the extrusion cooking process because of its low cost, sustainability, and versatility for production of a wide variety of food products such as expanded cereals, pasta, and instant meals [28]. This is a technology based on thermal processes of high-temperature-short time (HT-ST). Extrusion cooking can be defined as a continuous process in which materials, such as proteins and starches, are plasticized to form a fluid melt in a chamber or barrel as a result of high temperature, pressure, and shear stress, causing the material to be conveyed and forced to flow through a die of specific shape [29]. An extruder is composed of basic elements such as a barrel, single or twin screw, heating and cooling jackets, die, pressure recorder, raw material feeder, and controllers of screw speed and feed rate. Because extrusion cooking produces different types of food, extruders have become more specialized for food applications [30]. The extruders can be classified into two types: single screw and twin screw. A single-screw extruder was the first equipment used for food development for a direct cooking and forming application. It is mechanically simpler, less expensive, and easier to maintain but has some drawbacks in terms of operation, such as poor mixing and the necessity of premixing of ingredients and feeding conditioning before the process. Twin-screw extruders are classified according to the direction of rotation of the screws, in the same direction (corotating) or in the opposite directions (counterrotating), and the degree of intermeshing. Corotating twin-screw extruders are the most common in the food and snack industry for their efficiency, good control of residence time distribution, self-cleaning mechanism, and processing uniformity [31, 32]. These extruders are considerably more versatile than single-screw extruders and show more stable operation, with a wide range of applications and stability of product quality. The control of extrusion process operations—along with knowledge about the effects of variables of operation, such as temperature, screw speed, and feed moisture content—is necessary to obtain products with various desired physicochemical characteristics. Additionally, other inherent factors of extrusion equipment, such as the screw profile, size and shape of the die, and length and diameter of the barrel, are important geometrical characteristics that should be consistent with the characteristics of the desired food product.

During processing of materials, raw materials are conditioned (cleaning, classification, grinding, and conditioning to required moisture levels) and mixed with various ingredients such as bioactive compounds to produce diverse types of products of different shapes. Raw materials are fed into the extrusion equipment, where they are mixed and subjected to heating and friction. The solid phase is transformed into fluid melt at high temperature and pressure and forced to flow through the die. Due to the pressure change between barrel chamber and atmospheric pressure, instant vaporization occurs, and we get an expanded product with porous structure (**Figure 1**), which will depend on operating conditions and composition of the mix, among other factors. At the same time as structural changes occur in the solid matrix, e.g., starch gelatinization, protein denaturation and solubilization, and formation of complexes between amylose and lipids, there are reactions of degradation of antioxidants (such as vitamins, polyphenols, anthocyanins, and pigments), which are influenced by the type and intensity of the thermal and mechanical energy applied and are related to the process variables and screw and barrel geometric configurations.

2.1. Effects of extruder variables on product properties

Although the extrusion process is basically a simple technological operation, its control is complicated owing to the effects exerted by some variables of the process. The processing conditions are determined by independent and dependent variables of the system. The independent variables are those that can be controlled, such as feed composition, moisture content, rate of feed, screw speed, and barrel temperature. The dependent variables are those that assume a certain value that depends on the magnitude of an independent variable. These include the properties of extrudates, such as viscosity, which is affected by the composition, moisture content, temperature, and shear rate associated with the screw speed. The flow rate

Figure 1. Expansion of cornstarch due to extrusion.

is associated with configuration of the screw barrel, screw speed, viscosity, and pressure drops in the system; other properties that can be included here are pressure exerted on the system, power, specific energy, residence time, and product characteristics (texture, gelatinization, color, water absorption index, expansion index, density, and chemical composition, among others). Small changes in these variables can affect the quality and characteristics of the final product. The most influential variables in extrusion processes are temperature, screw speed, and system pressure. In addition, moisture content of the mixtures influences properties such as viscosity of the melt fluid, residence time of the material in the extruder, and shear stress applied to the food, thereby affecting the physical characteristics of extrudates or energy consumption. **Figure 2a** shows the effects of extrusion temperature and moisture content on a process designed to obtain an instant extruded cereal with the addition of apple pomace. It was found that specific mechanical energy input is affected by the moisture content of the mixtures, where higher values of specific energy are obtained at moisture content of 28%, and a further increase causes a decrease in energy input. This phenomenon can be attributed to the gelatinization mechanism, which is minimized at high

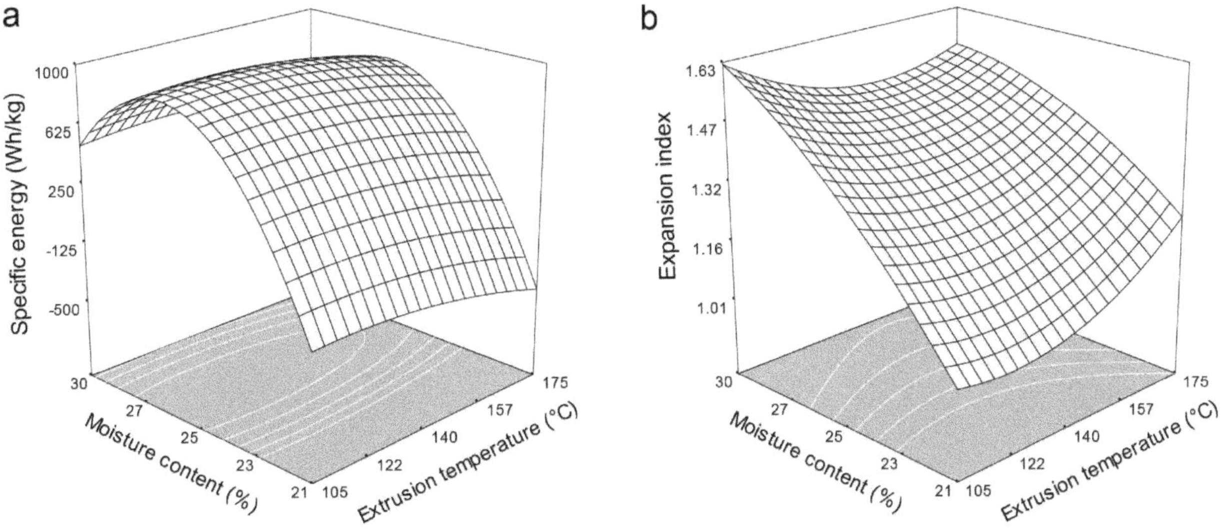

Figure 2. A response surface plot linking the effects of feed moisture content and extrusion temperature on specific energy (a) and the expansion index (b) of an extruded cereal with apple pomace.

moisture content, causing sliding of the material during the process. The expansion index (**Figure 2b**) is affected by the moisture content of the mixtures and extrusion temperature. Increases in feed moisture content increase expansion of the extrudates at low temperatures. Nonetheless, an increase in temperature reduces the expansion even at high moisture content. Evidently, properties such as expansion are dependent on the abovementioned factors, which correlate with the degree of gelatinization and composition of the mixtures [33]. Besides, the behavior of the expansion index could be correlated with fiber content of apple pomace, as reported in other studies, where it was found that fiber addition minimizes the expansion of cereals [8, 10].

Another study [34] has shown the effect exerted by the feed rate, resulting in a high expansion index, with a low water solubility index (WSI) and high hardness of extrudates. The increase in feed moisture content results in products with high density, low expansion, high water absorption index (WAI), lower WSI, and high hardness. The increase in barrel temperature increased expansion of the extrudates but reduced density with an increase in the WSI. These results show the effects of operating conditions during extrusion of cereals (feed rate, feed moisture content, screw speed, and barrel temperature) on physicochemical characteristics of the extruded products, e.g., on expansion, density, WAI, and WSI. These studies show versatility of the extrusion process for obtaining quality products with a suitable nutritional balance and functional characteristics that can be presented in various ways.

2.2. The extrusion cooking process for the development of functional foods

Some of the food extrusion applications are instant extruded products such as breakfast cereals, snack foods, baby food, instant soups, instant flour, and others. **Figure 3** shows a process diagram for the development of functional extruded products. Extruder machines are integrated into a production line of the extrudates.

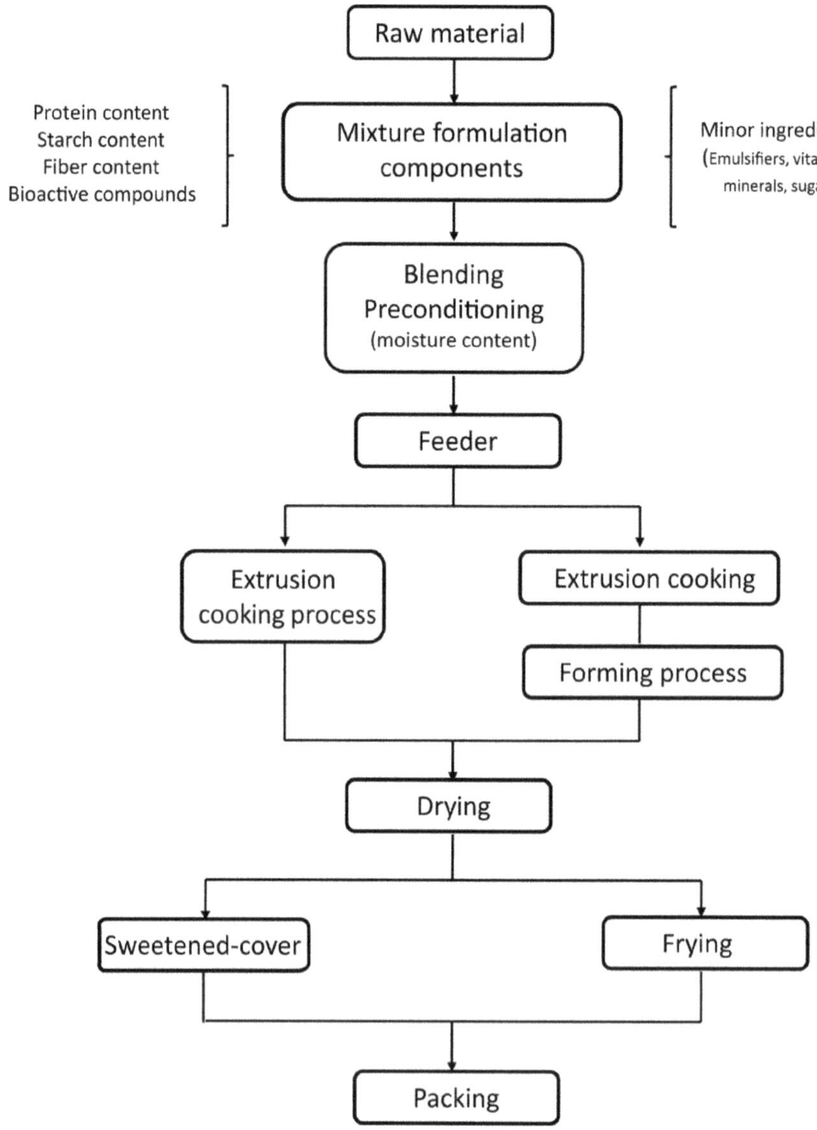

Figure 3. A flow diagram of the extrusion process for production of an expanded functional food.

The extrusion cooking process involves temperatures >100°C, where the food is mixed, transported, heated, and forced through one or more restricted openings (dies) at the discharge end of the barrel, expanding when emerging from the die. Subsequently, this extrudate is dried, sweetener coated, and packed, or directly packed. In the cold extrusion process, the processing temperature is low [35], and the product is mixed and formed without noticeable cooking, which typically causes degradation of some components. For this purpose, besides low temperatures, the screw and barrel configuration for low friction (a deep-flighted screw and smooth barrel) and low screw speeds are used. Some of the extruded products are described in **Table 3**. Among these are expanded cereals with different forms and appearances, extruded fruit products, extruded products based on proteins, and confectionery products. Each one can be reformulated for the production of functional extruded products via incorporation of soluble fiber, antioxidants (polyphenols, anthocyanins, or vitamins), or low-calorie sweeteners with the addition of calcium and microelements, proteins, and natural colorants, among others.

Characteristics	Products
Extruded cereal products	Ready-to-eat cereals Expanded snack foods Precooked flour Pasta products Pellets for snacks Bases for instant soups
Extruded fruit products	Fruit leather Fruit bars
Extruded protein products	Snacks (protein bars) Textured vegetable protein products Restructured seafood Semimoist and expanded pet foods Processed cheeses
Confectionery products	Chocolate Fruit gum Chewing gum

Heldman and Hartel [36]; Riaz [37].

Table 3. Food products obtained by an extrusion process.

3. Stability of functional components during the extrusion cooking process

Lately, the research in this field is focused on the development of functional foods because of the industry tendencies and consumer demands. A wide range of these kinds of products are obtained through an extrusion cooking process as described above. Nonetheless, during this process, chemical changes occur, affecting the presence and concentration of bioactive compounds in the extruded final product. For this reason, extruder equipment is considered a reactor of high-stress mechanical and thermal conditions that can accelerate chemical reactions among the components present in some ingredients, causing changes in their structures such as breaking of physical bonds, leading to the loss of functional compounds. The stability of bioactive or functional compounds in extruded products is reported as a loss or increase after extrusion cooking. **Table 4** shows the stability of some bioactive compounds used in the extrusion cooking process.

The losses of bioactive compounds are mainly due to the effects of temperature and mechanical stress changing the structure of compounds. A well-studied change is degradation of anthocyanins during this process. These could be converted to chalcones and small molecules without previous hydrolysis of the glycosidic bonds to form the corresponding aglycons [43, 44]. Anthocyanins that are more affected by high temperature are cyanidin 3-arabinoside and peonidin 3-arabinoside, in contrast to cyanidin 3-galactoside, cyanidin 3-glucoside, peonidin 3-galactoside, and peonidin 3-glucoside [6]. Other authors have reported degradation of cyanidin 3-glucoside into protocatechuic acid [11]. Regarding the reduction in the amounts of other polyphenols, the main change caused by the thermal process during extrusion is decarboxylation of free phenolic acids, which promotes polymerization of phenolics and tannins, thereby leading to reduced extractability and antioxidant activity [45].

Bioactive compounds losses				
Source	**Bioactive compounds**	**Process conditions**	**Loss (%)**	**Reference**
Thiamin	Thiamin	Moisture content: 11, 13, and 14% Temperature: 140–200°C Screw speed: 65, 73, and 81 rpm Residence time: 85–131 s Thiamin content: 9, 37, and 93 mg/kg	67–100%	[38]
Blueberry concentrate	Anthocyanins	Temperature: until 138°C Screw speed: 300 rpm	90%	[3]
Grape juice	Anthocyanins	Temperature: until 138°C Screw speed: 300 rpm	74%	[3]
Yellow maize grits	L-Lysine	Moisture content: 13, 15, and 17% Temperature: 150, 165, and 180°C Screw speed: 65, 73, and 81 rpm Residence time: 89–101 s	51–89%	[39]
Fruit powders (berries)	Anthocyanins	Temperature: 163°C Residence time: 3.5 min	65%	[4]
L-Acid ascorbic	L-Acid ascorbic	Temperature: 75–150°C Screw speed: 100 and 300 rpm Residence time: 31.58–48.81 s	56.4–79.2%	[40]
Navy beans	Phenolic compounds	Moisture content: 22% Temperature: 160°C Screw speed: 150 rpm Bean content: 15, 30, and 45%	10%	[10]
Small red beans	Phenolic compounds	Moisture content: 22% Temperature: 160°C Screw speed: 150 rpm Bean content: 15, 30, and 45	70%	[10]
Blueberry pomace	Anthocyanins	Moisture content: 45% Temperature: 160, 180, and 200°C Screw speed: 150 and 200 rpm	33–42%	[5]

Bioactive compounds losses

Source	Bioactive compounds	Process conditions	Loss (%)	Reference
Kiwicha	Polyphenols	Temperature: 180°C Screw speed: 254.5 rpm Residence time: 10–13 s	64.4–80.3%	[41]
Dried cranberry pomace	Procyanidins DP4 and DP9	Moisture content: 30% Temperature: 150, 170, and 190°C Screw speed: 150 and 200 rpm Pomace content: 30, 40, and 50%	DP4: 23–28% DP9: 68–77%	[6]
Bilberry extract	Anthocyanins	Moisture content: 22% Temperature: 100, 130, and 160°C Residence time: ≤ 60 s	90%	[42]
Chokeberry	Anthocyanins	Temperature: 100–140°C Screw speed: 300–500 L/min Chokeberry content: 1.2 kg/38.8 kg starch	35–100%	[43]
Blue maize	Anthocyanins	Moisture content: 30% Temperature: 80°C Extruder speed: 30 Hz	41–53%	[9]
Blue maize	Total polyphenols	Moisture content: 30% Temperature: 80°C Extruder speed: 30 Hz	12.5–38.3%	[9]
Pumpkin	α-Carotene	Moisture content: 15% Temperature: 150–170°C Pumpkin content: 4, 6, and 8%	100%	[21]
L-Acid ascorbic	L-Acid ascorbic	Moisture content: 15% Temperature: 150–170°C Pumpkin content: 4, 6, and 8%	HT: 49–76% LT: 13–40%	[21]
Black rice	Bond polyphenols	Moisture content: 12–17% Temperature: 60, 100, and 120°C Screw speed: 200 rpm	15%	[11]

Bioactive compounds losses				
Source	**Bioactive compounds**	**Process conditions**	**Loss (%)**	**Reference**
Red cactus pear encapsulated powder	Betacyanins	Moisture content: 0.22 kg/kg Temperature: 80, 100, 120, and 140°C Screw speed: 225, 275, and 325 rpm Residence time: 36.62–60.13 s Powder content: 2.5% (w/w)	33–51%	[2]
Red cactus pear encapsulated powder	Betaxanthins	Moisture content: 0.22 kg/kg Temperature: 80, 100, 120, and 140°C Screw speed: 225, 275, and 325 rpm Residence time: 36.62–60.13 s Powder content: 2.5% (w/w)	55–77%	[2]
Red cactus pear encapsulated powder	Polyphenols	Moisture content: 0.22 kg/kg Temperature: 80, 100, 120, and 140°C Screw speed: 225, 275, and 325 rpm Residence time: 36.62–60.13 s Powder content: 2.5% (w/w)	36–47%	[2]
Oat/apple pomace	Polyphenols	Moisture content: 21–30% Temperature: 104–175°C Screw speed: 150 and 200 rpm Oat/apple pomace content: 40/14%	2.9–20.1%	[1]
Bioactive compound increases				
Source	Bioactive compounds	Process conditions	Increase (%)	Reference
Bean/corn	Total polyphenols	Moisture content: 16.3% Temperature: 50–190°C Screw speed: 90 rpm	23%	[12]
Bean/corn	Flavonoids	Moisture content: 16.3% Temperature: 50–190°C Screw speed: 90 rpm	36%	[12]

Bioactive compounds losses				
Source	**Bioactive compounds**	**Process conditions**	**Loss (%)**	**Reference**
Blueberry pomace	Monomers biologically important	Moisture content: 45% Temperature: 160, 180, and 200°C Screw speed: 150 and 200 rpm	18%	[5]
Dried cranberry pomace	Flavonols (FL) Procyanidins DP1 and DP2	Moisture content: 30% Temperature: 150, 170, and 190°C Screw speed: 150 and 200 rpm Pomace content: 30, 40, and 50%	FL:30–34% DP1: 61–157% DP2: 49–164%	[6]
Black rice	Total polyphenols (TP) and free polyphenols (FP)	Moisture content: 12–17% Temperature: 60, 100, and 120°C Screw speed: 200 rpm	TP: 12.6% FP: 17%	[11]

Table 4. Stability of bioactive compounds used in the extrusion cooking process.

On the other hand, **Table 4** shows an increase in the concentration of some bioactive compounds mainly flavonols and procyanidins. The flavonols upregulated after the extrusion process are myricetin, quercetin, and their various glucosides. This change may be explained as follows: many flavonols may get bound to the cell wall components, especially after damage to cells [46]. Another explanation of the increased flavonol content is enhanced extraction of compounds because of disruption of the solid matrix upon extrusion [6]. Similar findings for other polyphenols such as caffeic acid and p-coumaric acid at some specific temperatures and moisture conditions have been reported [1]. Another change observed in polyphenols is the increased amounts of free polyphenols because extrusion leads to higher extraction efficiency and decreased amounts of bound polyphenols; researchers have also observed an increase in the amounts of procyanidins of low molecular weight (DP1 and DP2) as a result of decreased concentrations of procyanidins with DP $\geq$ 4 [6].

The antioxidant activity of extruded products is related to the amount of bioactive compounds and their losses caused by the extrusion cooking process. The general tendency is that the extrusion cooking process leads to losses of bioactive compounds and a decrease in the antioxidant activity as a consequence. Nonetheless, in some cases, antioxidant capacity is increased due to the structural changes in compounds. **Table 5** shows studies where antioxidant activity was determined before and after extrusion cooking, reporting a loss or increase under different process conditions and with different sources of bioactive compounds.

Some researchers have studied the changes in individual polyphenols using black rice [11] and apple pomace [1] as sources of bioactive compounds subjected to the extrusion cooking process. The stability of individual polyphenols is shown in **Table 6**.

Source	Bioactive compound	Extrusion cooking conditions	Antioxidant activity change (%)	Reference
Navy	Total polyphenols	Moisture content: 22% Temperature: 160°C Screw speed: 150 rpm Beans content: 15, 30, and 45%	Losses: 22%	[10]
Small red beans	Total polyphenols	Moisture content: 22% Temperature: 160°C Screw speed: 150 rpm Beans content: 15, 30, and 45%	Losses: 65%	[10]
Bean/corn	Polyphenols Flavonoids	Moisture content: 20% Temperature: 180°C	Increase: 27%	[12]
Kiwicha	Polyphenols	Temperature: 180°C Screw speed: 254.5 rpm Residence time: 10–13 s	Losses: 29–58%	[41]
Blue maize	Anthocyanins	Moisture content: 30/100 g Temperature: 80°C Extruder speed: 30 Hz	Losses: 12.8–34.9%	[9]
Red cactus pear encapsulated powder	Betalains	Moisture content: 0.22 kg/kg Temperature: 80, 100, 120, and 140°C Screw speed: 225, 275, and 325 rpm Residence time: 36.62–60.13 s Powder content: 2.5% (w/w)	Losses: 55–47%	[2]
Pumpkin	Lutein, zeaxanthin, and α-carotene	Moisture content: 15% Temperature: 150–170°C Pumpkin content: 4, 6, and 8%	Increase	[21]

Table 5. Losses and increases of antioxidant activity under the influence of the extrusion cooking process.

3.1. Modeling of degradation of bioactive compounds

Kinetic models of zero- and first-order reactions can predict the stability of these functional compounds. Thermomechanical degradation of functional components during an extrusion cooking process has been mostly fitted to first-order models, such as the one in Eq. (1):

Polyphenol	Source	
	Black rice*	Oat/apple pomace**
Gallic acid	45% (⇓)	–
Syringic acid	58.7% (⇑)	–
Chlorogenic acid	27.1% (⇑)	57–71% (⇓)
Caffeic acid	39.5% (⇓)	55–64% (⇓)
p-Coumaric acid	15.7% (⇑)	38–51% (⇓)
Ferulic acid	13.7% (⇑)	25–28% (⇓)
Rutin	–	56–70% (⇓)
Phloridzin	–	46–76% (⇓)
Epicatechin	–	0% (⇓)

⇑ increase and ⇓ loss.
Ti et al. [11].
Leyva-Corral et al. [1].

Table 6. Stability of individual polyphenols in extruded products.

$$\mathrm{Ln}(C_t - C_0) = -kt \tag{1}$$

where C_0 represents the initial concentration of a component, C_t is the concentration of the component at time t, k denotes the reaction rate constant, and t is the residence time in the extruder. Furthermore, dependence of the constant reaction rate on the extrusion cooking temperature can be described by the Arrhenius equation, as shown in Eq. (2):

$$k = A_0 \exp\left(\frac{E_a}{RT}\right) \tag{2}$$

where A_0 is the pre-exponential, E_a is activation energy of the reaction, R is the universal gas constant, and T is absolute temperature.

Eq. (1) has been used to describe a decrease in the amounts of thiamin [38], amino acids [39], anthocyanins [42], and betalains [2], and Eq. (2) has been used to calculate activation energy during an extrusion cooking process. **Table 7** shows reported values of the degradation rate constant and activation energy. Studies have shown that thiamin degradation is dependent on extrusion temperature, feed moisture, and screw speed, but the degradation of thiamin is most dependent on extrusion temperature [38].

As for losses of amino acids—lysine, arginine, and cysteine—the results have shown that the reaction rate constants are strongly dependent on temperature, lysine being the amino acid that is more sensitive to temperature than the others are, although shear stress additionally affects the amino acid loss [39]. In natural pigments such as betacyanins (red-purple pigments) and betaxanthins (yellow-orange pigments), k values tend to decrease as

Bioactive compound	Kinetic parameters		
	k (s^{-1})	E_a (kJ/mol)	Reference
Thiamin	0.011–0.093	52.1	[38]
Lysine	0.012–0.073	127 ± 23	[39]
Arginine	0.0007–0.0019	68 ± 10	[39]
Cysteine	0.0012–0.0032	76 ± 24	[39]
Anthocyanins	0.2×10^{-4}–1.2×10^{-3}	45.33	[42]
Betacyanins	0.0188–0.0206	1.5888	[2]
Betaxanthins	0.0122–0.0167	6.1815	[2]

Table 7. Kinetic parameters of degradation of bioactive compounds during the extrusion cooking process.

temperature increased. An increase in temperature causes the material to flow faster in the extruder, and consequently, the pigments have shorter exposure to high temperature and shear stress during the extrusion. Furthermore, the rate constants for the degradation of betacyanins are higher than those obtained for betaxanthins. Nonetheless, activation energy for the degradation of betaxanthins was found to be greater than that for degradation of betacyanins, indicating that betaxanthins are more sensitive to a temperature increase during the extrusion cooking process [2]. For anthocyanins [42], both k values and activation energy are affected by temperature, but mechanical stressors and moisture of the material also have to be considered as parameters influencing anthocyanin degradation in an extrusion cooking process.

Author details

Martha G. Ruiz-Gutiérrez, Miguel Á. Sánchez-Madrigal and Armando Quintero-Ramos*

*Address all correspondence to: aquinter@uach.mx

Faculty of Chemical Sciences, Autonomous University of Chihuahua, Nuevo Campus Universitario Circuito Universitario s/n, Chihuahua, Chihuahua, México

References

[1] Leyva-Corral J, Quintero-Ramos A, Camacho-Dávila A, Zazueta-Morales JJ, Aguilar-Palazuelos E, Ruiz-Gutiérrez MG, Meléndez-Pizarro CO, Ruiz-Anchondo TJ. Polyphenolic compound stability and antioxidant capacity of apple pomace in an extruded cereal. LWT - Food Science and Technology. 2016;**65**:228-236. DOI:10.1016/j.lwt.2015.07.073

[2] Ruiz-Gutiérrez MG, Amaya-Guerra CA, Quintero-Ramos A, Pérez-Carrillo E, Ruiz-Anchondo TJ, Báez-González JG, Meléndez-Pizarro CO. Effect of extrusion cooking on bioactive compounds in encapsulated red cactus pear powder. Molecules. 2015;**20**:8875-8892. DOI: 10.3390/molecules20058875

[3] Camire ME, Chaovanalikit A, Dougherty MP, Briggs JL. Blueberry and grape anthocyanins as breakfast cereal colorants. Journal of Food Science. 2002;**67**:438-441. DOI: 10.1111/j.1365-2621.2002.tb11425.x

[4] Camire M, Dougherty M, Briggs J. Functionality of fruit powders in extruded corn breakfast cereals. Food Chemistry. 2007;**101**:765-770. DOI: 10.1016/j.foodchem.2006.02.031

[5] Khanal RC, Howard LR, Brownmiller CR, Prior RL. Influence of extrusion processing on procyanidin composition and total anthocyanin contents of blueberry pomace. Journal of Food Science. 2009;**74**:52-58. DOI: 10.1111/j.1750-3841.2009.01063.x

[6] White BL, Howard LR, Prior RL. Polyphenolic composition and antioxidant capacity of extruded cranberry pomace. Journal of Agriculture Food Chemistry. 2010;**58**:4037-4042. DOI: 10.1021/jf902838b

[7] Puupponen-Pimia R, Hakkinen ST, Aarni M, Suortti T, Lampi AM, Eurola M, Piironen V, Nuutila AM, Oksman-Caldentey KM. Blanching and long-term freezing affect various bioactive compounds of vegetables in different ways. Journal of the Science of Food and Agriculture. 2003;**83**:1389-1402. DOI: 10.1002/jsfa.1589

[8] Altan A, McCarthy KL, Maskan M. Evaluation of snack foods from barley-tomato pomace blends by extrusion processing. Journal of Food Engineering. 2008;**84**:231-242. DOI: 10.1016/j.jfoodeng.2007.05.014

[9] Sánchez-Madrigal MÁ, Quintero-Ramos A, Martínez-Bustos F, Meléndez-Pizarro CO, Ruiz-Gutiérrez MG, Camacho-Dávila A, Torres-Chávez PI, Ramírez-Wong B. Effect of different calcium sources on the bioactive compounds stability of extruded and nixtamalized blue maize flours. Journal of Food Science and Technology. 2015;**52**:2701-2710. DOI: 10.1007/s13197-014-1307-9

[10] Anton A, Fulcher RG, Arntfield SD. Physical and nutritional impact of fortification of corn starch-based extruded snacks with common bean (*Phaseolus vulgaris* L.) flour: Effects of bean addition and extrusion cooking. Food Chemistry. 2009;**113**:989-996. DOI: 10.1016/j.foodchem.2008.08.050

[11] Ti H, Zhang R, Zhang M, Wei Z, Chi J, Deng Y, Zhang, Y. Effect of extrusion on phytochemical profiles in milled fractions of black rice. Food Chemistry. 2015;**178**:186-194. DOI: 10.1016/j.foodchem.2015.01.087

[12] Delgado-Licón E, Martínez-Ayala AL, Rocha-Guzman NE, Gallegos-Infante JA, Atienzo-Lazos M, Drzewiecki J, Martínez-Sánchez CE, Gorinstein S. Influence of extrusion on the bioactive compounds and the antioxidant capacity of the bean/corn mixtures. International Journal of Food Sciences and Nutrition. 2009;**60**:522-532. DOI: 10.1080/0963748080197666

[13] Sharma, SK, Bansal, S, Mangal, M, Dixit, AK, Gupta, RK, Mangal AK. Utilization of food processing by-products as dietary, functional, and novel fiber: A review. Critical Reviews in Food Science and Nutrition. 2016;**56**:1647-1661. DOI: 10.1080/10408398.2013.794327

[14] Saguy I, Kopelman IJ, Mizrahi S. Thermal kinetic degradation of betanin and betalamic acid. Journal of Agricultural and Food Chemistry. 1978;**26**:360-362. DOI: 10.1021/jf60216a052

[15] Huang AS, von Elbe JH. Kinetics of the degradation and regeneration of betanine. Journal of Food Science. 1985;**50**:1115-1120. DOI: 10.1111/j.1365-2621.1985.tb13024.x

[16] Herbach KM, Stintzing FC, Carle R. Thermal degradation of betacyanins in juices from purple pitaya [*Hylocereus polyrhizus* (Weber) Britton & Rose] monitored by high-performance liquid chromatography-tandem mass spectrometric analyses. European Food Research and Technology. 2004;**219**:377-385. DOI: 10.1007/s00217-004-0948-8

[17] Dlamini NR, Taylor JRN, Rooney LW. The effect of sorghum type and processing on the antioxidant properties of African sorghum-based foods. Food Chemistry. 2007;**105**:1412-1419. DOI: 10.1016/j.foodchem.2007.05.017

[18] Altan A, McCarthy KL, Maskan M. Effect of extrusion process on antioxidant activity, total phenolics and β-glucan content of extrudates developed from barley-fruit and vegetable by-products. International Journal of Food Science and Technology. 2009;**44**:1263-1271. DOI: 10.1111/j.1365-2621.2009.01956.x

[19] Devi PB, Vijayabharathi R, Sathyabama S, Malleshi NG, Priyadarisini VB. Health benefits of finger millet (*Eleusine coracana L.*) polyphenols and dietary fiber: A review. Journal of Food Science and Technology. 2014;**51**:1021-1040. DOI: 10.1007/s13197-011-0584-9

[20] Escalante-Aburto A, Ramírez-Wong B, Torres-Chávez PI, López-Cervantes J, Figueroa-Cárdenas JD, Barrón-Hoyos, JM, Morales-Rosas I, Ponce-García N, Gutiérrez-Dorado R. Obtaining ready-to-eat blue corn expanded snacks with anthocyanins using an extrusion process and response surface methodology. Molecules. 2014;**19**:21066-21084. DOI: 10.3390/molecules191221066

[21] Obradović V, Babić J, Šubarić D, Jozinović A, Ačkar D, Klarić I. Influence of dried Hokkaido pumpkin and ascorbic acid addition on chemical properties and colour of corn extrudates. Food Chemistry. 2015;**183**:136-143. DOI: 10.1016/j.foodchem.2015.03.045

[22] Brouillard R. Chemical structure of anthocyanins. In: Markakis P, editor. Anthocyanins as Food Colors. New York: Academic Press Inc; 1982. pp. 1-40

[23] He J, Giusti MM. Anthocyanins: Natural colorants with health-promoting properties. Annual Review of Food Science and Technology. 2010;**1**:163-187. DOI: 10.1146/annurev.food.080708.100754

[24] Tsuda T. Dietary anthocyanin-rich plants: Biochemical basis and recent progress in health benefits studies. Molecular Nutrition and Food Research. 2012;**56**:159-170. DOI: 10.1002/mnfr.201100526

[25] Sánchez-Madrigal MÁ, Quintero-Ramos A, Martínez-Bustos F, Meléndez-Pizarro CO, Ruiz-Gutiérrez MG. Effect of different calcium sources on the antioxidant stability of tortilla chips from extruded and nixtamalized blue corn (*Zea mays L.*) flours. Food Science and Technology. 2014;**34**:143-149. DOI: 10.1590/S0101-20612014000100021

[26] Zielinski H, Michalska A, Piskula MK, Kozlowska H. Antioxidants in thermally treated buckwheat groats. Molecular Nutrition & Food Research. 2006;**50**:824-832. DOI: 10.1002/mnfr.200500258

[27] Reyes LF, Villarreal JE, Cisneros-Zevallos L. The increase in antioxidant capacity after wounding depends on the type of fruit or vegetable tissue. Food Chemistry. 2007;**101**:1254-1262. DOI: 10.1016/j.foodchem.2006.03.032

[28] Rossen JL, Miller RC. Food extrusion. Food Technology. 1973;**27**:46-53

[29] Harper JM, Clark JP. Food extrusion. C R C Critical Reviews in Food Science and Nutrition. 1979;**112**:155-215. DOI: 10.1080/10408397909527262

[30] Camire, ME. Extrusion cooking. In: Henry CJK, Chapman C, editors. The Nutrition Handbook for Food Processors. 1st ed. Boca Raton FL: Woodhead Publishing Ltd; 2002. pp. 314-330

[31] Schuler EW. Twin-screw extrusion cooking system for food processing. Cereal Food World. 1986;**31**:413-416

[32] Muthukumarappan K, Karunanithy Ch. Extrusion process design. In: Ahmed J, Rahman MS, editors. Handbook of Food Process Design. Oxford, UK: Wiley-Blackwell; 2012. pp. 710-742. DOI: 10.1002/9781444398274.ch25

[33] Chinnaswamy R, Hanna MA. Physicochemical and macromolecular properties of starch-cellulose fiber extrudates. Food Structure. 1991;**10**:229-239

[34] Ding Q, Ainsworth P, Tucker G, Marson H. The effect of extrusion conditions on the physicochemical properties and sensory characteristics of rice-based expanded snacks. Journal of Food Engineering. 2005;**66**. pp. 283-289

[35] Guy RCE, Horne AW. Extrusion and co-extrusion of cereals. In: Blanshard JMV, Mitchell JR, editors. Food Structure–Its Creation and Evaluation. Ist ed. London: Butterworth; 1988. pp. 331-350. ch18

[36] Heldman DR, Hartel RW. Principles of Food Processing. New York: Chapman and Hall; 1997. pp. 253-283

[37] Riaz MN. Selecting the right extruder. In: Guy R, editor. Extrusion Cooking-Technologies and Applications. 1st ed. Cambridge: Woodhead Publishing; 2001. pp. 29-50

[38] Ilo S, Berghofer E. Kinetics of thermomechanical destruction of thiamin during extrusion cooking. Journal of Food Science. 1998;**63**:312-316. DOI:10.1111/j.1365-2621.1998.tb15732.x

[39] Ilo S, Berghofer E. Kinetics of lysine and other amino acids loss during extrusion cooking of maize grits. Journal of Food Science. 2003;**68**:496-502. DOI: 10.1111/j.1365-2621.2003.tb05701.x

[40] Plunkett A, Ainsworth P. The influence of barrel temperature and screw speed on the retention of L-ascorbic acid in an extruded rice based snack product. Journal of Food Engineering. 2007;**78**:1127-1133. DOI: 10.1016/j.foodeng.2005.12.023

[41] Repo-Carrasco-Valencia R, Pena J, Kallio H, Salminen S. Dietary fiber and other functional components in two varieties of crude and extruded kiwicha (*Amaranthus caudatus*). Journal of Cereal Science. 2009;**49**:219-224. DOI: 10.1016/j.jcs.2008.10.003

[42] Hirth M, Leiter A, Beck SM, Schuchmann HP. Effect of extrusion cooking process parameters on the retention of bilberry anthocyanins in starch based food. Journal of Food Engineering. 2014a;**125**:139-146. DOI: 10.1016/j.foodeng.2013.10.034

[43] Hirth M, Preiß R, Mayer-Miebach E, Schuchmann HP. Influence of HTST extrusion cooking process parameters on the stability of anthocyanins, procyanidins and hydroxycinnamic acids as the main bioactive chokeberry polyphenols. LWT - Food Science and Technology. 2014b;**62**:511-516. DOI: 10.1016/j.lwt.2014.08.032

[44] Jackman RL, Yada RY, Tung MA, Speers RA. Anthocyanins as food colorants: A Review. Journal of Food Biochemistry. 1987;**11**:201-247. DOI: 10.1111/j.1745-4514.1987.tb00123.x

[45] Brennan C, Brennan M, Derbyshire E, Tiwari BK. Effects of extrusion on the polyphenols, vitamins and antioxidant activity of foods. Trends Food Science and Technology. 2011;**22**:570-575. DOI: 10.1016/J.tifs.2011.05.007

[46] Hutzler P, Fischbach R, Heller W, Jungblut T, Reuber S, Schmitz R, Veit M, Weissenbock G, Schnitzler J. Tissue localization of phenolic compounds in plants by confocal laser scanning microscopy. Journal of Experimental Botany. 1998;**49**:953-965. DOI: 10.1093/jxb/49.323.953

7

Indirect Extrusion: A Multifaceted Approach of Sub-surface Tubular Expansion

Tasneem Pervez, Sayyad Z. Qamar,
Omar S.A. Al-Abri and Rashid Khan

Abstract

Extrusion and indirect extrusion is a very old manufacturing process used in multitudes of applications mainly focused on transportation, household and power industries. Indirect extrusion has found an interesting application in petroleum industry, which resulted in resolving many unsolvable issues over the last few decades. The current and expected future global demand for hydrocarbons became a driving force for researchers to find new comprehensive and cheaper solutions for hydrocarbon production. The challenges faced in oil and gas fields, while drilling, constructing and operating new and old vertical/horizontal wells, are many. The use of indirect extrusion for insitu expansion of sub-surface tubulars used in wells revolutionized the drilling and completion as opposed to one and half decade back. The emergence of solid expandable tubular technology has changed the basics of how we design and construct wells. The original development of the technology was to overcome the challenges faced by the petroleum industry to reach ultra-deep reservoirs, off-shore drilling, drilling in high-pressure/difficult zones and repair/maintenance of old/ageing wells. However, it gained significant interest of researchers and operators in providing solutions to wide-range problems. The development of a computational framework using finite element method (FEM) enabled to determine the force required for expansion and resulting dimensional changes in final product, which is of direct assistance to the field engineers. The effect of friction and stress variations along contact surface is also determined.

Keywords: indirect extrusion, tubular expansion, oil and gas wells, finite element method, solid expandable tubular, nonlinear analysis

1. Introduction

Extrusion is the process by which a block/billet of metal is reduced in cross section by forcing it to flow through a die orifice under high pressure. It is classified as: (a) Direct Extrusion, in which the ram of the machinery moves and forces the billet (semifinished metal form) through the die and (b) Indirect Extrusion, in which the ram is stationary and the die moves forcing the billet through the die. The process keeps the die stationary using a stem that is longer than the container containing the billet. Pipe extrusion is another process by which various forms of pipes are manufactured using a broad range of polymers selected based on intended usage. Pipes made of aluminum are also manufactured using extrusion. However, an interesting form of indirect extrusion occurs, when ferrous metal pipes are *in situ* expanded by forcing the die to pass through it. This novel use of pipe manufacturing, when extended for subsurface applications results in expanded pipe and is termed as solid expandable tubular (SET). The use of solid expandable tubular has made huge strides in the way the oil and gas wells are drilled and completed. Not only this, it has proven that the technique is able to provide solution to many unresolved problems faced by the petroleum industry.

Solid expandable tubular technology is a recent development in petroleum exploration and drilling. Essentially, it is a down-hole process consisting of expanding the diameter of the tubular by pushing or pulling a cone through it. Recent extensive research and development work in expandable tubular has provided successful solutions for a variety of difficult problems such as zonal isolation, aged fields, deteriorated casings, difficult-to-access and uneconomical reservoirs, deep drilling, etc. It has also helped to move forward toward the long-desired target of monodiameter wells. Still under development, but has proved its worth to the energy sector, especially due to its economic and environmental benefits. It aims at reducing the incurred and operational cost near 20–30%. These issues are not only long-standing but also have far-reaching consequences in petroleum industry as it involves the industry's most fundamental technologies; wellbore tubular.

Viewed from an engineering mechanics point of view, the tubular expansion is strongly nonlinear due to geometric and material nonlinearities as well as the contact conditions between the tubular and cone. These characteristics pose challenges to obtain viable and accurate solutions. A mathematical model is developed to predict the force required for expansion, variations in length and thickness, surplus deformation, the stress and strain patterns in expanded tubular, and the effect of material velocities in pre and post expansion regions. The governing equations are solved analytically and using finite element method. The use of finite element method is needed to determine the effect of material velocity during expansion. In order to expand the tubular, the required force is determined for various shapes of the mandrel and the cone associated with mandrel shape. The variation in required expansion force also varies with variation in friction coefficients and expansion ratios (ERs). There is an increase in expansion force with an increase in these two parameters. During tubular expansion process, the tubular material above and below the expansion zone gets effected, which was quantified through magnitude of plastic deformation, reduction in tubular wall thickness, and material velocity in pre and post expansion zones. It was found that for lower values of mandrel cone angles and expansion ratios greater than 24%, the tubular fails even at very low friction coefficients. The maximum interfacial stress (contact stress) occurs at 20% expansion ratio while displaying an upward trend with increase in coefficient of friction between the mandrel and the tubular surfaces. Simulation results asserted that the cone, during its progress, defines the new diameter of the tubular and found out that the surplus deformation or the gap between the cone diameter and the expanded tubular inner diameter is negligible. A comparative study is also done between the analytical and simulation results.

2. Literature review

Metal forming operations are classified as processes that are designed to exploit a remarkable property of metal plasticity, the ability to flow as solid without deterioration of properties. The simple rearrangement of the work piece to produce the desired shape and size reduces the waste substantially. To impose such alterations, external forces with appropriate boundary conditions are applied. Tubular expansion, an indirect extrusion method, is an important manufacturing process, which is used in the production of a wide range of products ranging from house hold pipes to boilers and heat exchanger tubes [1, 2]. In recent years, the expansion of circular tubes has drawn more and more attention in much more critical applications where accurate dimensions, smooth surface finishes, and long-term safe and trouble-free operations are of great importance such as catheters for medical application, passive energy absorbers in automotive industry, etc. [3–5].

At the juncture of 20th and 21st centuries, the development of an innovative application of tube expansion in oil and gas industry led to the solid expandable tubular (SET) technology. Multitudes of successful research work was done on concept, significances and applications of solid expandable tubular technology; however, the focus was mostly on industry demand-driven needs. Filippov et al. [6], in his pioneering work, presented three solution scenarios using SET in drilling. First, an expandable open-hole drill liner that can reduce costs and enhance the ability of the operator to access new reservoirs, which are currently difficult to drill economically. Second, expandable cased-hole liner to maintain profitable production from older fields and third is an expandable liner hanger system that can reduce or eliminate the occurrence of the hydraulic leaking of liners, which can improve the economics of drilling deeper and farther.

In SET, the deformation is strongly nonlinear due to large strains, complex material behavior, and contact conditions between the tubular and the mandrel. These characteristics make it quite difficult to obtain exact solution. Decades of research work on cold-work tube forming produced different analytical or semi-analytical models with simplifying assumptions to determine force required to permanently enlarge tube inner diameter, resulting changes in its length and thickness, material strain hardening history, etc. The pioneering work of Hill [7] for tube sinking was a major breakthrough in determining an analytical or semi-analytical solution. However, very limited research work is done on developing analytical and simulation models of tubular expansion process, including inherent nonlinearities, its dynamics, the effect of interaction between the tubular and surrounding fluids, contact mechanics, etc. [8, 9]. Another challenge was accurate prediction of the tubular expansion in irregular shaped boreholes. The boreholes are mostly not concentric and circular in cross section. This results in oval-shaped expansion of tubular, which does not allow enough clearance for completion tools to pass through. The simulation of such expansion was successfully done using finite element method and design graphs were generated for field engineers to make right decision before lowering the completion to complete an oil well [10]. By combining SET with sidetracking technology, the operators mitigated troublesome shale formations and reached additional reserves with sufficient hole size. In summary, expandable tubular can thus be considered as a potential innovative breakthrough. Although the solid expandable tubular technology provided solutions for many unresolved problems, but limitations remain due to its increased cost, long-term sustainability, and issues related to health, safety and environment. The major obstacle comes from lack of thorough understanding of tubular material and corresponding mechanical strength, once the inner diameter of the tubular is increased beyond 25% of its original value. The prediction of post expansion mechanical properties is necessary to avoid any failure during its service life.

There were very limited attempts that tackled the task of developing a comprehensive analytical model for the tubular expansion process, and mostly under very simplified conditions. For instance, Marciniak and Duncan [11] developed a model for tubular expansion under tension by assuming a constant wall thickness and neglecting the axial strain within the expansion zone. Using a very simplified definition, Stewart et al. [12] considered the tubular expansion process to be similar to a burst test stopped at a pressure between yield pressure and burst rupture pressure. Thus, with aid of the basic definitions of stress and strain along with some empirical equations for burst pressure, very simplified formulations were developed for some of the basic parameters of tubular expansion process like expansion ratio, equivalent plastic strain, burst pressure, and expansion force. The mathematical model developed by Fischer et al. [13] for tube flaring considered the variation in pressure and thickness within the expansion zone and determined the expansion force required to expand thin walled tubes in conical form under compression. Yeh [14] conducted a theoretical study to determine the relationship between punch stroke, tube thickness, expansion ratio, and flaring limit. He found that material strain hardening has no effect on the relationship between thickness variation and expansion ratio. Through relating the assumptions of Marciniak and Duncan [11], Al-Hiddabi et al. [15] used plasticity and membrane theories to develop models for expansion process of thin-walled tubular with a conical expansion tool. The models demonstrated the variation in the force required for expansion with respect to expansion ratio, friction coefficient (μ), mandrel geometry, and tubular material's yield strength. Later, Ruan and Maurer [16] developed another model for tubular expansion process based on force and energy balance principles. This model is capable of estimating the contact pressure at the tubular-mandrel interface, as well as predicting the amount of pressure needed to drive the mandrel forward in order to attain the desired expansion ratio. With an aim of optimizing the mandrel shape, Karrech and Seibi [17] came up with another model for expansion of thin-walled tubular with an aim to estimate stress value, expansion force, and dissipated energy as a result of the expansion process.

First successful attempt to develop a comprehensive analytical model was done by Al-Abri et al. [18], which have reduced significant number of assumptions of previous models. In addition, the results of analytical model fare very well with computational and experimental results in terms of all parameters of interest including the contact pressure between the mandrel and tubular contact zones. For tubulars with threaded connections, a nonuniform contact condition occurs at the mandrel-tubular interface. During expansion of these threaded sections of the tubular, the disengagement of threads results in additional load on adjacent threads, which in turn damages it badly resulting in lowering the dependability of threaded connections [19]. The root cause of such behavior is the axial bending, which can be significantly reduced by lowering the mandrel cone angle from a typical value of 10 to 2°. None of these attempts do properly account for all the effects associated with down-hole expansion process, which demands that the tubular must expand to the desired diameter without fracturing, bursting, or damaging the tubular. At the same time, it must have hydraulic capabilities to provide sufficient resistance against burst and collapse during service. Due to the difficulty in obtaining closed form solutions of the problem in its entirety, researchers opted for approximate solutions using computational methods [20]. The study showed the effects of expansion ratio, friction coefficient, and cone angle on tubular expansion. Even at low values of friction coefficient, the failure could not be avoided for small mandrel cone angles and large expansion ratios. Researchers have tried aluminum tubular due to its improved formability as compared to steel, but the severity of down-hole environment excludes this possibility [21, 22].

In order to analyze the stress, strain, and residual stress induced in the solid expandable tubular applied in repairing casing damaged in wells, Binggui et al. [23] used the commercial finite element method (FEM) package ANSYS to model and simulate the problem. The expansion process induces residual stress in the tubular after expansion. Using the effective measures of the residual stress obtained through the developed model, a proactive action can be taken to reduce the residual stress in tubular through properly designed manufacturing or postprocessing stages. Another important observation drawn from the study was the occurrence of secondary deformation in the SET during the use of high-pressure liquid; it was found that the radius size unilaterally increased by 0.9–1.3 mm. A careful study of published literature shows that most of the developed models for tubular expansion have ignored the dynamic effects by simply assuming that the expansion process occurs out at a very low speed. It is true that under this assumption strain rate effects are negligible but not the dynamic friction conditions, which originates at the interfaces [24].

Right from the inception of SET, a long-term goal has been set for gradually elimination of the conventional telescopic well through the development of slim and monodiameter wells [25–27]. With the help of slim wells, the rate of penetration can be improved, optimal operation of solids control equipment and can achieve lower capacity rigs. Similarly, Matthew [28] defined the challenges and potential of a monodiameter well. The anticipation that expandable monodiameter well will provide the operators an ability to isolate zones that contain reactive shale's, subsalt environments, low fracture gradient formation, or other drilling situations without reducing the casing sizes, is a significant benefit. Due to this, one can reach the reservoir with adequate production hole sections at total depth, hence will avoid adverse effect on oil production. Ruggier et al. [29] have documented the advances in expandable tubing for zonal isolation and casing repair. The main advantage of using expandable tubing as opposed to conventional means is that it can restore the casing to its original integrity without a significant loss in internal diameter. One can also apply SET in fixing the aquifer casing leak with substantial savings [30]. Carlos et al. [31] and Jennings [32] reported the application of SET in reducing the tapering effect that occurs when using multiple casing strings in deepwater drilling. These expandable systems conserve valuable hole size at total depth. Conserving hole size, or diametric efficiency (DE), allows additional pipe strings to be deployed to drill deeper or farther. Marketz et al. [33] described the application of SET in carbonate reservoirs. The technology, when used on trial basis in more than 20 wells in fractured carbonate reservoirs, demonstrated success with expandable tubular solutions to replace scab liners for fracture shutoff and to replace cement for zonal isolation. In addition, fracture shutoff in the drilling phase with expandable tubular and swelling elastomer has made it possible to reduce water cuts of horizontal wells in a carbonate reservoir.

3. Oil well tubular expansion

The principle behind solid expandable tubular technology in an oil well is simple. The tubular expands by push or pull of the mandrel through the tubular. First, a mandrel is housed in a small section of the tubular and welded to the unexpanded tubular, thus creating a string of expanded and unexpanded sections of the tubular. The mandrel can be pushed forward by using high-pressure fluid at the rear end of the mandrel or can be pulled by using a rod, which is connected to the front end of the mandrel. Either of these moves the mandrel forward resulting in expansion of the tubular. **Figure 1** shows the expanded and unexpanded sections of the tubular along with the mandrel. Due to the forward motion of the mandrel, under application of external force, the tubular deforms from initial inner diameter to the required inner diameter. It is extremely important to maintain the structural integrity of the tubular during the expansion process. The expanded tubular must be able to provide sufficient resistance against burst and collapse during its lifetime operation. Furthermore, critical data such as wall thickness, length changes, postexpansion strength, ductility, burst, and collapse strengths

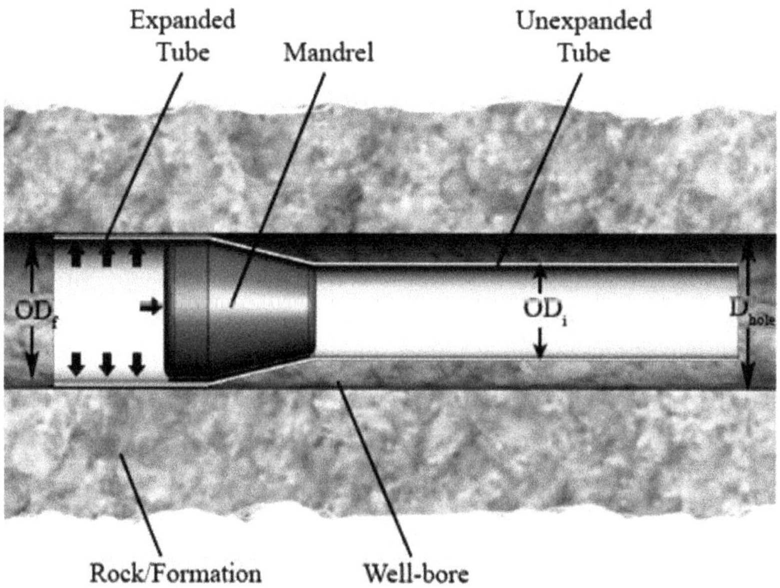

Figure 1. Tubular expansion in an oil well.

are essential for its safe and reliable use. During the expansion process, high intensity of stress at the mandrel-tubular interface necessitates careful design of mandrel and selection of its material. In particular, the mandrel geometry becomes important to successfully execute the expansion process.

Ideally, an expanded tubular should have constant diameter and wall thickness over the entire length of the expanded section. Due to the circular cross section of mandrel, it is expected that the expanded tubular cross section will also be circular. Various laboratory and field trials have shown that the expanded sections along the depth remain circular in cross section but their centers may not necessary lie on the same axis. However, if the borehole is of irregular shape, then restricted space for expansion of the tubular results in complex boundary conditions, which result in noncircular cross section of tubular. Expansions in the range of 8–26% of the initial tubular inner diameter have been successfully achieved during full-scale laboratory tests but in field applications, expansions of up to 20% have been successfully attained.

4. Finite element formulation

When studying the deformation of metallic solids due to applied loads, it is common to assume that the material will behave in an isotropic manner. Typically, this assumption is extended to both the elastic and elastic-plastic responses. The forming process may induce anisotropy into the material, giving directionally sensitive elastic-plastic mechanical behavior. Nevertheless, in tube forming (tubular expansion), the induced anisotropy due to permanent deformation is assumed negligible. Huber-Mises yield criterion is used to determine onset of yielding and is updated during the material strain hardening history. Contact between mating surfaces is defined using Coulomb friction law. An associated flow rule, in which the plastic flow potential is assumed to be the yield function, is used to define the elastic-plastic incremental constitutive relation. By means of incremental and iterative procedure, the numerical solution is obtained including expansion force, thickness and length changes, and surplus deformation. The law defining the onset of yielding under combined state is known as the yield criterion. With the assumption of no Bauschinger effect, the final form of yield function, $f(\hat{\sigma})$, as it is required for current analysis is given by

$$f(\widehat{\sigma}) = \boldsymbol{\sigma} - S_Y = 0 \tag{1}$$

where $\hat{\sigma}$ is the second order stress tensor, $\boldsymbol{\sigma}$ is the von-Mises or effective stress and S_Y is the yield strength of the tubular material. Using the plastic potential function, $g(\hat{\sigma})$, we can define a flow rule. In the present work, an associated flow rule is assumed, so that the yield and the plastic potential surfaces coincide, i.e., $f = g$. Hence, the plastic flow develops along the normal to the yield surface. The phenomenon whereby yield stress increases with further plastic straining is known as strain hardening. In strain hardening theory of plasticity, one must relate the hardening parameters to the experimental stress-strain curve. A stress variable called effective stress has been defined above. In order to define a corresponding strain variable, called effective plastic strain, which is a function of the plastic strains, the concept of plastic work, i.e., work conjugate relation, is used. Using simple algebraic manipulation, the final equation for effective plastic strain increment $d\varepsilon_p$, in terms of plastic strain components $d\hat{\varepsilon}_p$ can be written as

$$d\varepsilon_p = \sqrt{(2/3)d\hat{\varepsilon}_p \cdot d\hat{\varepsilon}_p} \tag{2}$$

Once the material yields, its behavior will be partly elastic and partly plastic. During any incremental change of loading, the change in strain is assumed to be decomposed additively into elastic and plastic components, i.e.,

$$d\epsilon = d\epsilon_e + d\epsilon_p \tag{3}$$

The elastic strain increment is related to stress increment through an elastic stress/strain matrix "D" [34]. Using the consistency condition, which requires the state of stress to remain on the yield surface during plastic flow, one can derive the following incremental stress/strain relationship:

$$d\hat{\sigma} = \left[D - \left\{D\frac{\partial f}{\partial\hat{\sigma}} * D\frac{\partial f}{\partial\hat{\sigma}}\right\}\left\{E_p + D\frac{\partial f}{\partial\hat{\sigma}} * D\frac{\partial f}{\partial\hat{\sigma}}\right\}^{-1}\right]d\hat{\varepsilon} \tag{4}$$

where the term in brackets represents incremental plastic stiffness matrix [K]. The incremental elastic-plastic stress/strain relation is valid only when there is plastic deformation. Thus before using this relation, we need loading criteria to determine whether the material is in a state of plastic deformation or not. The loading criteria can be obtained using consistency condition, which assures that the stress state remains on the yield surface as given in Eq. (1).

$$\begin{aligned} &\frac{\partial f}{\partial\hat{\sigma}} \bullet d\hat{\sigma} > 0: \quad \textit{Loading} \\ &\frac{\partial f}{\partial\hat{\sigma}} \bullet d\hat{\sigma} = 0: \quad \textit{Neutral loading} \\ &\frac{\partial f}{\partial\hat{\sigma}} \bullet d\hat{\sigma} < 0: \quad \textit{Unloading} \end{aligned} \tag{5}$$

If the material is not in a state of plastic deformation, the elastic stress/strain relation given by Eq. (6) should be used instead of Eq. (4)

$$d\hat{\sigma} = Dd\hat{\varepsilon} \tag{6}$$

Considering pressure loads "p" and traction forces "t" between surfaces, the incremental principle of virtual work gives:

$$\int_{Volume} \{\delta\varepsilon_i\}^T\{\sigma_i\}\,dV = \int_{Area} p\,\delta u\,dA + \int_{Surface} \delta u\,t\,dS \tag{7}$$

The contact between mandrel and tubular was modeled using Coulomb friction law to account for the induced friction between the interacting surfaces. The general procedure for defining a finite element model from the principle of virtual work is well known and is clearly stated in literature [34]. The final assembled equation for finite element analysis is given below.

$$[K]\,\{du\} = \{F_{Pressure}\} + \{F_{Contact}\} \quad (8)$$

where [K] and $\{du\}$ are the assembled stiffness matrix and displacements at grid points. Modified Newton-Raphson method is used to obtain the solution. The strain rate effect is neglected in the above formulation due to low mandrel velocity. The rigid body mandrel was set to travel along the axis of the tubular at a low speed of 1.5 m/min to expand the tubular. Due to low mandrel speed, the strain rate effect was assumed to be negligible.

5. Simulation of tubular expansion in vertical well

Finite element method (FEM) is used to simulate both, axisymmetric and 3-dimensional (3-D) models of the expansion process. The finite element formulation described in Section 4 represents a more generalized approach, which is typical in obtaining simulation results using FEM. The formulation can be programmed using a generic computer language but requires extensive effort to get graphical output from simulation results. On the other hand, various commercial finite element programs are available, which use similar approaches and have excellent postprocessing features. Among a host of available commercial software, ABAQUS fits very closely to the above formulation and hence is used for simulation purpose instead of writing own code.

The tubular was modeled as a deformable body with elastic-plastic material behavior; whereas the mandrel was modeled as a rigid body. The mandrel was also constrained from rotation about its own axis. The induced friction between the tubular and mandrel was modeled using Coulomb friction law. The mandrel cone angle was varied from 10 to 45°. Force required for expansion, equivalent plastic strain, effective stress, contact force, thickness and length variations, and surplus deformation were extracted from output data file of simulated cases. **Figure 2(a)** shows the three-dimensional finite element model of the tubular and mandrel. Due to symmetry only one half of the tubular was considered. The tubular was modeled using eight-node brick elements with reduced integration. After trial runs using 2D and 3D models, a 2D axisymmetric model was selected as shown in **Figure 2(b)** due to cost and computational time considerations. The edges of the mandrel were fillet of 6 mm radius to avoid stress concentration. The tubular was held fixed at expanded end and kept free on unexpanded end. This allowed tubular expansion under tension. Expansion ratios of 5–35% were considered for simulation.

ASTM standard test method (ASTM E8) is used to do tests on three tubular specimens. The specimens cut from the tubular material were prepared according the instructions stated in ASTM standard. Tinius Olson universal testing machine is used to conduct tensile test until

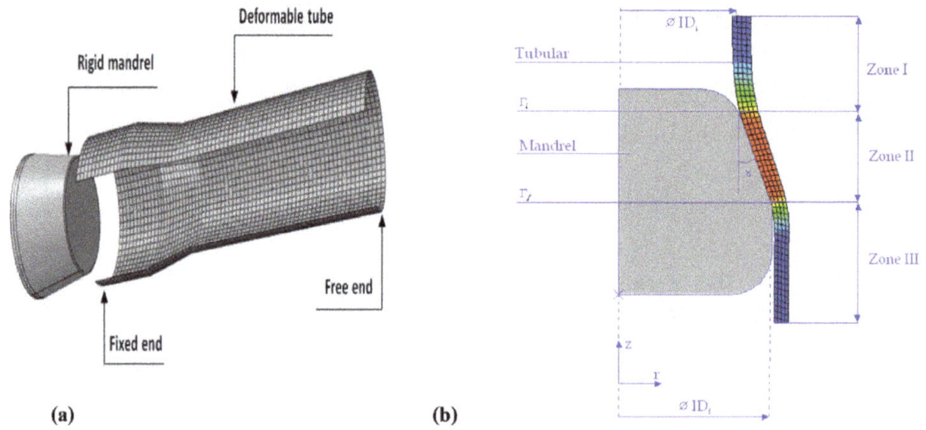

Figure 2. (a) 3-D finite element model; (b) 2-D axisymmetric model with pre- and postexpansion zones.

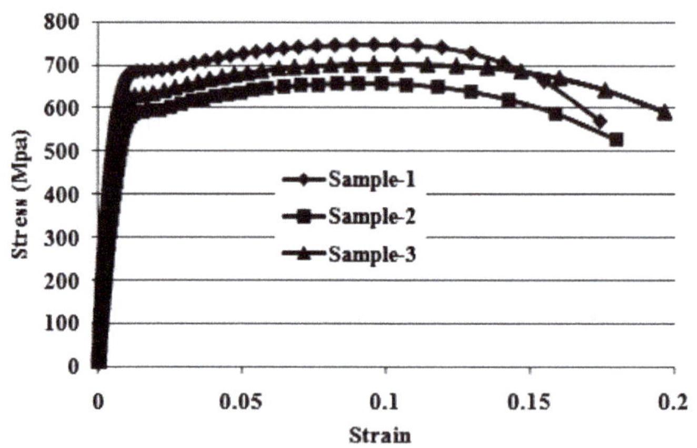

Figure 3. Stress-strain curve for tubular material under uniaxial tensile test.

the specimen fractures. The average of three samples, as shown in **Figure 3**, is then used to determine Young's modulus, yield strength, ultimate tensile strength, strength at fracture, and ductility.

For simulation, the material properties used are the average of these samples. These average stress-strain data are provided as a direct input to the ABAQUS software. The tubular dimensions and other input parameters of finite element model are given in **Table 1**.

Each simulation run consists of two distinct steps. First, it allowed the inner surface of the tubular to expand by prescribing radial displacements up to the final expansion ratio. Then, it released the displacements to establish free equilibrium to define the equilibrium under no load. The stability of the solution is achieved by carefully selecting step sizes. Several factors do affect the solid tubular expansion process; however, the two important parameters for well engineering applications are friction coefficient (μ) and expansion ratio (ER), which were investigated with respect to expansion ratio. Variation in these two parameters will result in an increase in required expansion force. Higher expansion force will lead to two scenarios; (a) need to have larger oil-rig platforms, or (b) need to lower the expansion ratio or friction coefficient, which will reduce the final tubular diameter and hence the target depth of the

Part	Parameters	Value
Tubular (geometry)	Inner diameter	174.625 mm
	Outer diameter	193.675 mm
	Wall thickness	9.525 mm
	Section length	500.000 mm
Tubular (material)	Young modulus of elasticity	210.00 GPa
	Yield strength	615.00 MPa
	Ultimate tensile strength	702.50 MPa
	Poisson's ratio	0.3
	Strain at fracture	0.1966
Mandrel/tubular	Coefficient of friction	0.1/0.2/0.3/0.4
Mandrel	Mandrel cone angle	10°/20°/45°

Table 1. Finite element model input parameters.

reservoirs. The term expansion ratio (ER) is defined as the ratio of difference between the mandrel and the preexpanded tubular diameter to preexpanded tubular diameter.

5.1. Expansion force and stress

Figure 4 shows the stress contours of effective stress for four different expansion ratios at constant values of μ = 0.4 and mandrel cone angle of 20°. Although the stress contours look similar indicating that the expanded zone reaches yielding for all cases, tubular wall thickness decreases with an increasing expansion ratio as discussed in later part of this section. A prior knowledge of expansion requirements is necessary to select appropriate expansion tools and avoid any unexpected failure. The drawing force required for expansion of different tubular sizes is extremely important. Finite element analysis of tubular expansion was carried out to determine expansion force for different expansion ratios, mandrel cone angles and friction at tubular/mandrel interface as shown in **Figures 5** and **6**. **Figure 5** shows the expansion force variation as a function of mandrel position for 25% expansion ratio and different friction coefficients. The maximum force occurs at the beginning of expansion process to overcome inertia, which later drops to almost constant magnitude during rest of the expansion. The small fluctuations in force are due to the transients but overall the expansion process is stable. The average value of expansion force increases to two folds, when μ changes from 0.1 to 0.4. Similar variations in expansion force were found from low expansion ratio (5%) to high expansion ratio (35%) for μ = 0.1 (**Figure 6(a)**). In this case, the increase in average expansion force for 35% ER is more than three time than what is needed for 5% ER. The tubular wall thickness reduces significantly for large expansion ratio and smaller mandrel cone angle. The major part of thickness reduction occurs due to large expansion ratio, which conforms well to volume constancy condition. However, a separate study conducted by authors showed that small cone angle also results in thickness reduction but not to the extent due to large expansion ratios [18]. This phenomenon can be avoided by using a spherical shape mandrel instead of

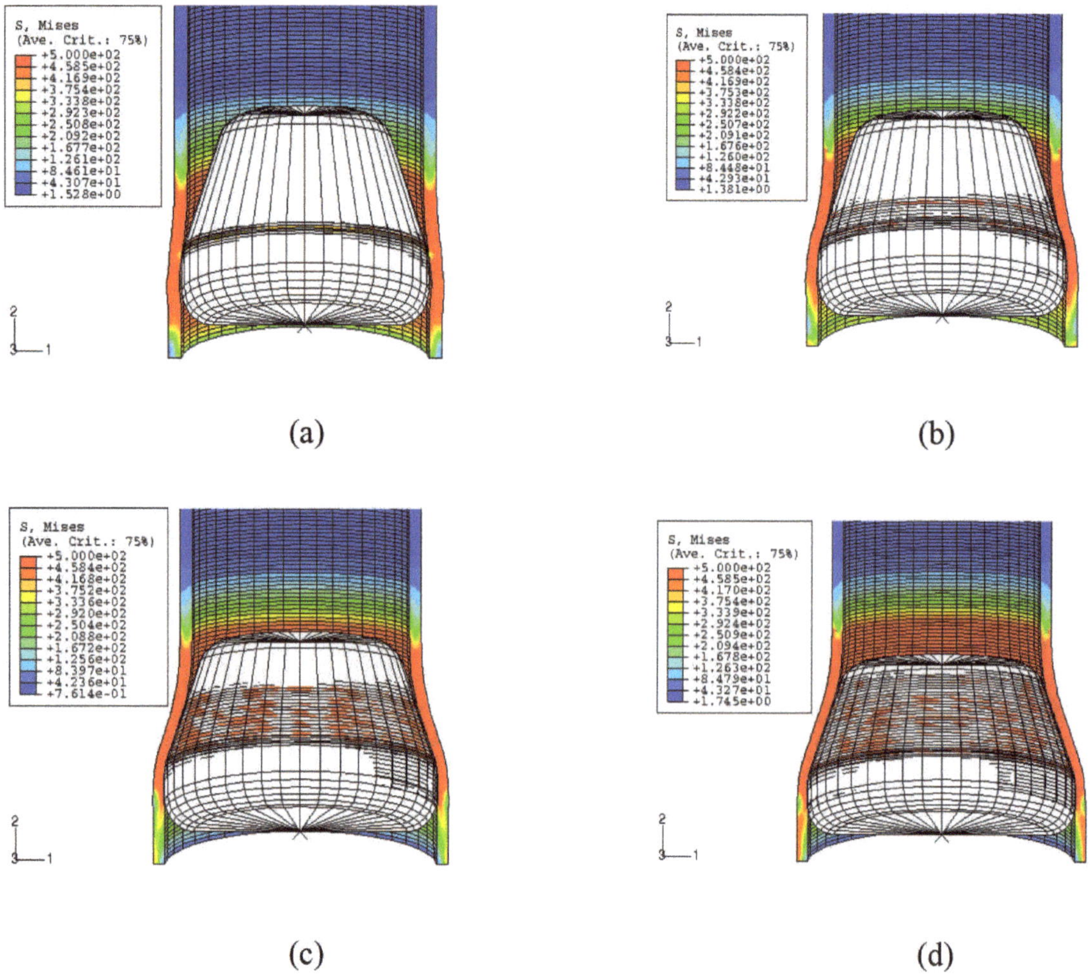

Figure 4. Contours of effective stress, in MPa, for $\mu = 0.4$ and mandrel cone angle of 20°, (a) 5%; (b) 15%; (c) 25% and (d) 35% expansion ratios.

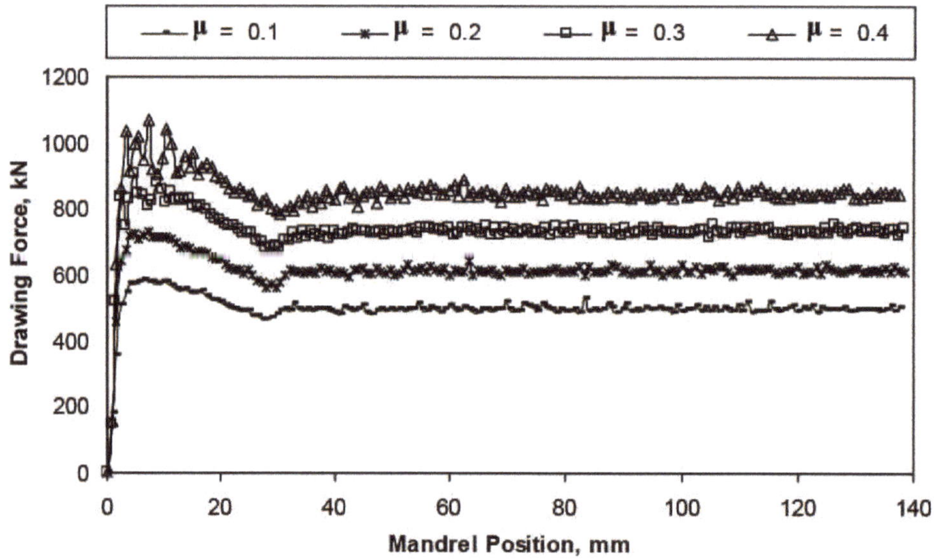

Figure 5. Variation in expansion force w.r.t mandrel position for 25% expansion ratio.

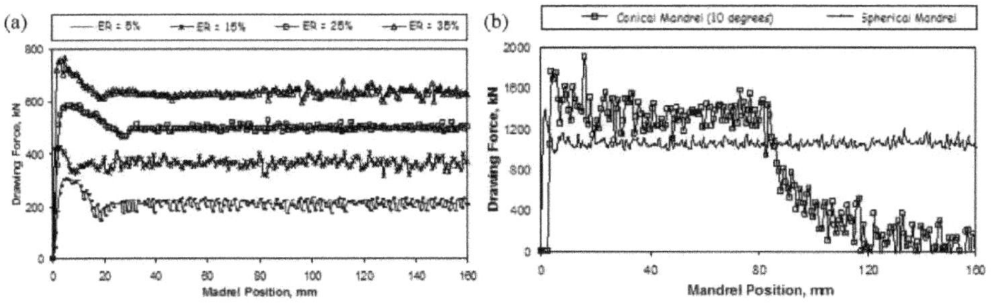

Figure 6. (a) Variation in expansion force w.r.t mandrel position for friction coefficient of 0.1; (b) variation in expansion w. r.t mandrel position for different mandrel shape (μ = 0.4).

conical shape. **Figure 6(b)** shows the variation in expansion force for conical and spherical mandrels for 35% ER and μ = 0.4. The tubular mandrel system remains stable for spherical mandrel and desired expansion is attained successfully. This is true for expansion under tension.

5.2. Material velocity

The velocity variation describes the material flow during the expansion process. To analyze this, divide the tubular into three material zones: I, II, and III as shown in **Figure 2(a)**. Zone I represents the unexpanded section of the tubular, which lies in front of the mandrel. The section of the tubular under expansion is termed as zone II, which is in direct contact with the mandrel. Zone III lies behind the mandrel and represents the expanded section of the tubular. **Figure 7(a) and (b)** show the variation in radial and axial velocity components of the tubular along tubular radius in zones I, II, and III. It is clear from **Figure 7(a)** that there is no change in the radial velocity component, V_r, along the tubular wall thickness, whether it is at zone I or II or III. Contrary to this, the axial velocity component, V_z, remains constant in zone II (z = 237 mm) but varies in other two zones, i.e., in front and back of the mandrel. It decreases in front of the mandrel (z = 217 mm) while increases at the back of the mandrel (z = 257 mm). The reason for this variation is the bending deformation mode of the tubular, which results in alternate tension and compression of inner and outer fibers of the tubular along its wall thickness. To explain it further, the outer and inner fibers of zone I are in compression and tension, respectively. This means that the outer and inner fibers move in the same or opposite direction to the mandrel's motion, which yield higher values of axial velocity in outer fibers as compared to the inner fibers, as shown in **Figure** 7.

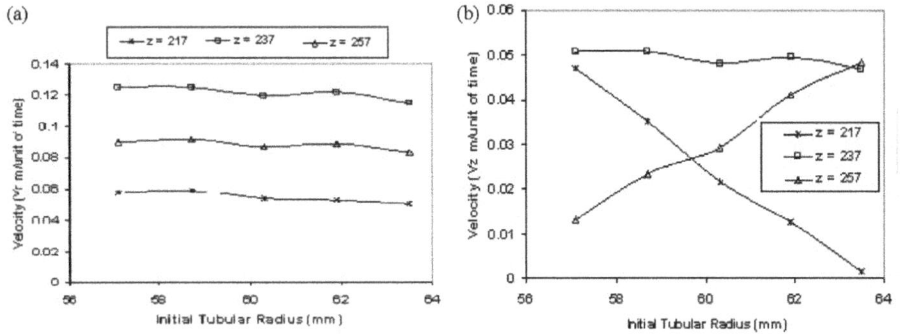

Figure 7. (a) Variation in radial velocity w.r.t tubular radius at different positions along z-axis; (b) variation in axial velocity w.r.t tubular radius at different positions along z-axis.

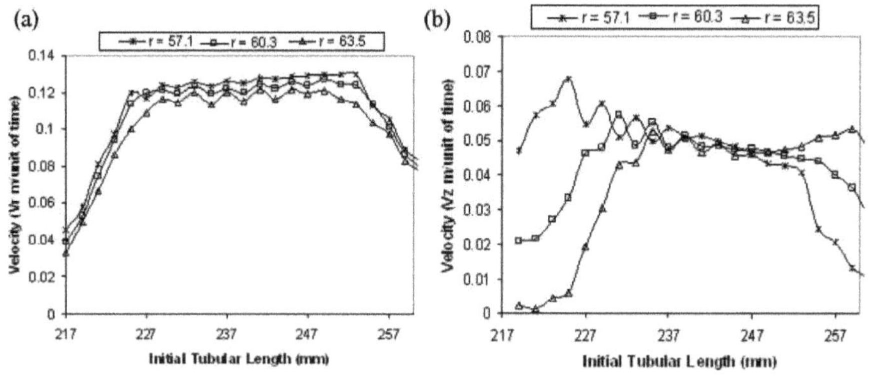

Figure 8. (a) Variation in radial velocity w.r.t tubular axis at different positions along its radius; (b) variation in axial velocity w.r.t tubular axis at different positions along its radius.

The variations in radial and axial velocities along tubular axis at different radial positions are shown in **Figure 8(a) and (b)**. There is an increase in radial and axial velocities with respect to the tubular axis in front of the mandrel, i.e., in zone I and they reduce to zero in the expanded section, i.e., zone III. As we move away from the two ends, only rigid body motion happens between the mandrel and the tubular along the tubular axis. According to the volume constancy, velocities V_z^i and V_z^f (with respect to r) of the surfaces Γ_i and Γ_f, respectively, are related by the following expression:

$$\frac{V_z^f}{V_z^i} = \frac{OD_i^2 - ID_i^2}{OD_f^2 - ID_f^2} \tag{9}$$

Figure 8(b) shows that the velocity in zone II decreases by 20% (0.05–0.04), which is same as predicted by above equation. Therefore, the model accurately predicts material flow along tubular axis. It is worth to mention that the ratio V_z^f/V_z^i decreases with the expansion ratio and increases with the thickness reduction.

5.3. Equivalent plastic strain

The prime reason for tubular expansion is to increase the inner diameter of the tubular. This takes place due to the elastic-plastic deformation of the tubular under applied load. The magnitude of the plastic deformation contributes to the enlargement of the inner diameter and its cumulative effect in terms of strain is termed as equivalent plastic strain. The variation in equivalent plastic strain ε_{eq}^p with respect to radial and axial directions are shown in **Figure 9 (a) and (b)**, respectively. As it can be seen from **Figure 9(a)**, the variation in ε_{eq}^p is negligible along the thickness of the tubular with a maximum in zone III. However, **Figure 9(b)** shows that ε_{eq}^p decreases along the z-direction.

Figures 10(a)–12(b) show the variation in plastic strain components with respect to radial and axial directions satisfying the volume constancy as given below:

$$\dot{\varepsilon}_{rr} + \dot{\varepsilon}_{\theta\theta} + \dot{\varepsilon}_{zz} = 0 \tag{10}$$

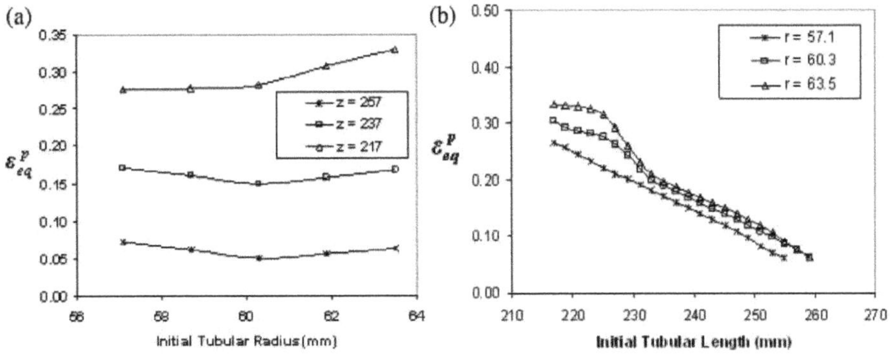

Figure 9. (a) Variation in equivalent plastic strain w.r.t tubular radius at different z-axis values; (b) variation in equivalent plastic strain w.r.t tubular axis at different radial positions.

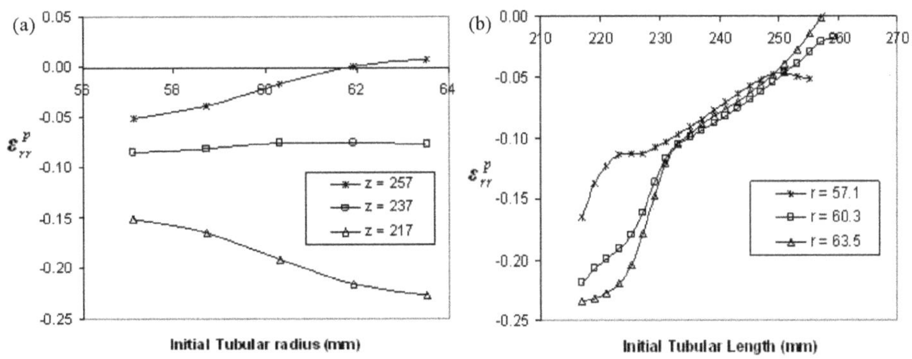

Figure 10. (a) Variation in radial plastic strain w.r.t to tubular radius at different "z" values; (b) variation in radial plastic strain w.r.t tubular axis at different radius.

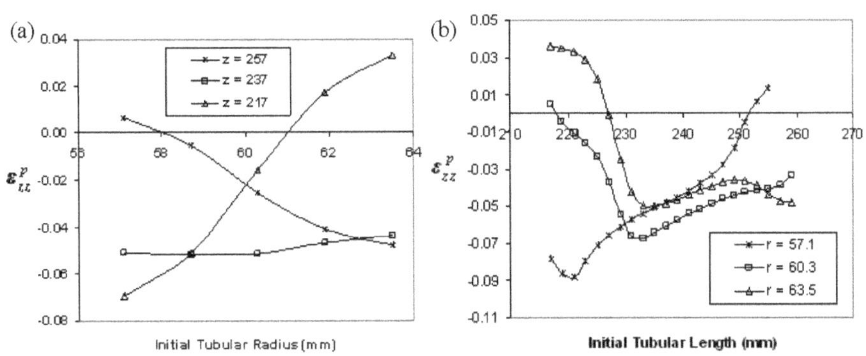

Figure 11. (a) Variation in axial plastic strain w.r.t tubular radius at different positions "z" values; (b) variation in axial plastic strain w.r.t tubular axis at different radius.

Assuming linear rate and neglecting the elastic deformation the components of the plastic strain must satisfy the following equation:

$$\varepsilon^p_{rr} + \varepsilon^p_{\theta\theta} + \varepsilon^p_{zz} = 0 \tag{11}$$

One can verify from **Figures 10(b)**, **11(b)**, and **12(b)** that at any arbitrary point (r = 51.7 mm and z = 257 mm), the components of strain sums to zero, i.e.,

$$\varepsilon^p_{rr} + \varepsilon^p_{zz} + \varepsilon^p_{\theta\theta} = -0.17 - 0.08 + 0.25 = 0 \tag{12}$$

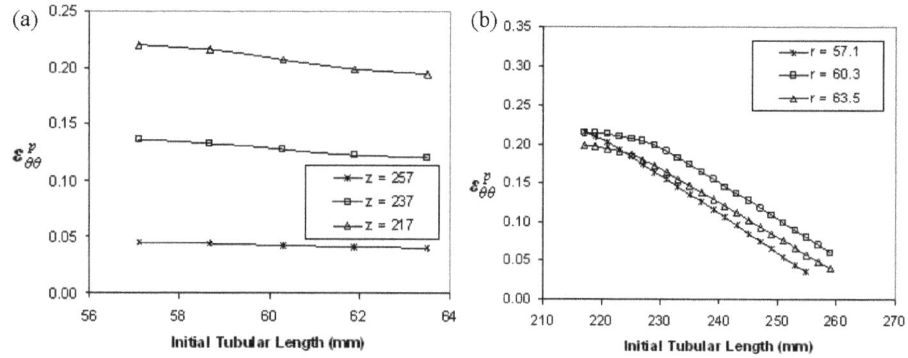

Figure 12. (a) Variation in circumferential plastic strain w.r.t tubular radius at different "z" values; (b) variation in circumferential plastic strain w.r.t tubular axis at different radius.

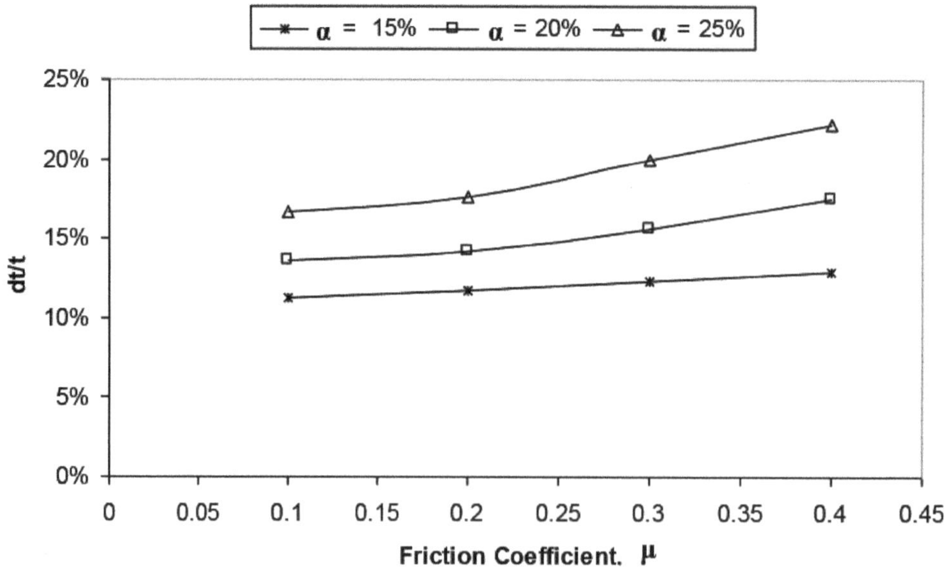

Figure 13. Thickness variation versus coefficient of friction for different expansion ratios.

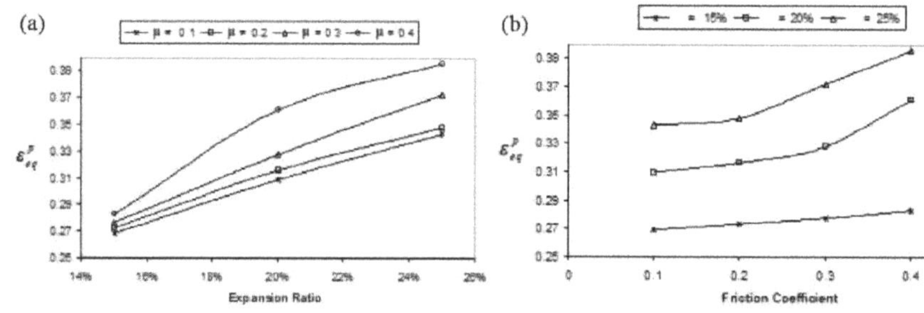

Figure 14. (a) Equivalent plastic strain versus expansion ratios for different friction; (b) equivalent plastic strain versus coefficient of friction for different expansion ratios.

The thickness reduction reached to as high as 25% for an expansion ratio of 25% and friction coefficient of 0.4 as can be seen from **Figure 13**. It also supports the findings of the result shown in **Figure 10(b)**. The reduction in wall thickness of the tubular increases with an increase in coefficient of friction and expansion ratio. This magnitude of reduction will have significant impact on postexpansion strength of the tubular particularly its burst and collapse strengths [18]. **Figure 14(a)** shows that for a fixed value of expansion ratio, the magnitude of equivalent plastic strain increases with the increase of friction coefficient. However, **Figure 14(b)** shows that for a fixed friction coefficient, the magnitude of equivalent plastic strain increases with expansion ratio. Therefore, it can be concluded that equivalent plastic strain ε_{eq}^{p} increases with friction coefficient and expansion ratio.

5.4. Contact stress

The contact condition between the mandrel and the tubular is of particular interest in tubular expansion. The forward motion of mandrel inside the tubular during expansion results in contact between the inclined surface of mandrel and zone II of the tubular. However, the simulation results show that the contact happens only at small areas at the beginning and end of zone II. The tubular section between the two ends of zone II does not come in contact with the mandrel. To understand this, let's define the normal velocity of the tubular with respect to the mandrel at the mandrel-tubular interface as given below:

$$V_n = V_r \cos\gamma - V_z \sin\gamma \tag{13}$$

where γ is the mandrel angle, which is equal to 20° between the two ends of zone II. However, at the two ends, γ is dependent on fillet radius. If magnitudes of velocities from **Figures 7(a)** to **8(b)** are substituted in above equation, we obtain nonzero velocity V_n in the middle of zone II, which means that there is no contact between the mandrel and the tubular at that particular region. However, full tubular-mandrel contact happens at the front and rear end of the mandrel. From mechanics point of view, it will only be possible, when the material in contact with the mandrel surface at the beginning of zone II is under tension (bottom side) and at the end of zone II is under compression (top side). This represents the bending deformation, which causes the outward deflection along the radial direction. The results of **Figure 15** justifie the presence of gap or no contact between the two ends. The two peaks of stresses are at the two ends of mandrel-tubular interface, while the stress value is zero in between, i.e., in the region of no contact between the tubular and the mandrel.

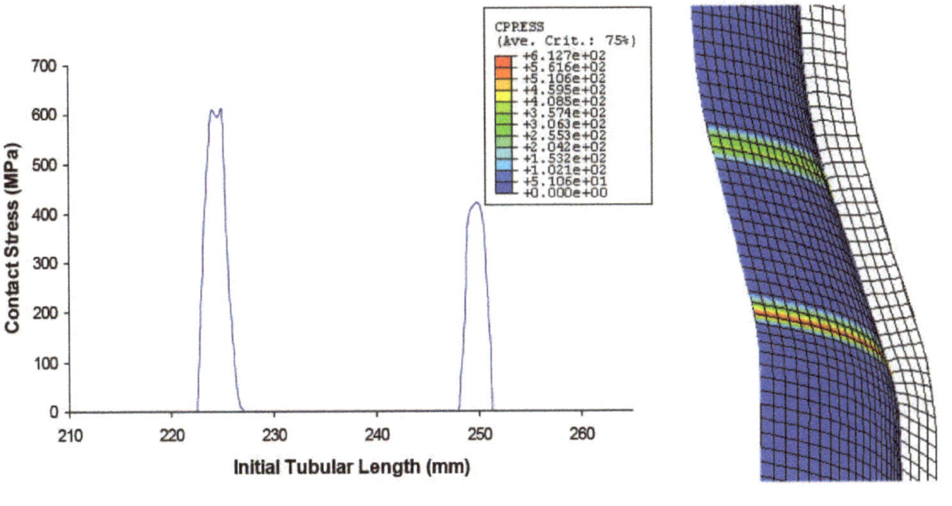

Figure 15. Variation in contact stress, in MPa, at mandrel-tubular interface.

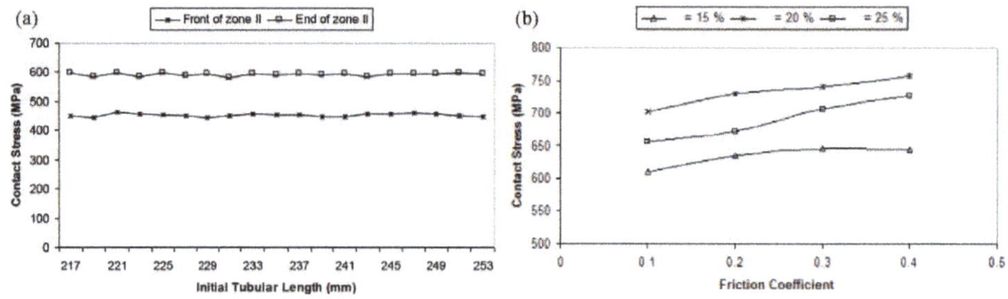

Figure 16. (a) Variation in contact stress, in MPa, at the beginning and end of zone II; (b) back end maximum contact stress versus friction coefficient for different ER.

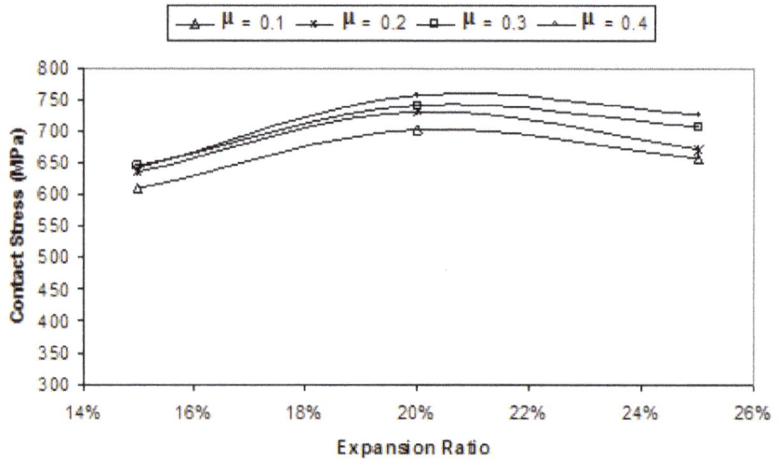

Figure 17. Back end maximum contact stress, in MPa, versus expansion ratio for different μ.

Figure 16(b) shows that for a fixed expansion ratio the maximum contact stress increases with the friction coefficient. However, for a fixed friction coefficient the maximum contact stress occurs at an expansion ratio of 20% as shown in **Figure 17**.

5.5. Effect of friction coefficient

Studying the effects of friction coefficient on expansion force, tubular length and thickness variation as well as the surplus was performed by keeping the mandrel angle and the expansion ratio constant while varying the friction coefficient. This effect was also conducted for various expansion ratios while maintaining the mandrel angle fixed for all cases. Previously published work presented the variation of the drawing force, tubular thickness and length, and surplus variation as a function of the mandrel position, which concluded that the drawing force increases as the friction coefficient increases for all cases regardless of the expansion ratio and the mandrel angle [18]. The force variation peeks at the start-up of the expansion process and decreases to a constant level with small fluctuation as the mandrel advances further indicating that expansion is becoming more stable. It was observed that the tubular thickness reduces as the friction coefficient increases. This reduction keeps increasing to a certain level and stays constant as the mandrel passes the points of interest where the results were extracted. This behavior is the same for all expansion ratios (5–35%) at the mandrel angles of 10 and 20°. However, for a mandrel angle of 45°, the tubular thickness reduction was observed to increase to a certain level and reduces to show some recovery in the order of 20% of the maximum thinning. The tubular length was observed to shorten and elongates for certain cases. In most of the cases, the tubular length shortens for small values of drawing forces as well as mandrel angles. However, the length elongates for higher drawing forces causing the expanded section to elongate further due to tension. Therefore, it can be concluded that the variation in tubular length depends heavily on the magnitude of the drawing force, which in turns depends on the tubular expansion, resistive force due to friction between metal surfaces, mandrel cone angle, and bending.

5.6. Effect of expansion ratio

Similar observations to the drag effect were made when studying the effect of expansion ratio on expansion force, tubular length and thickness, and tubular surplus while maintaining the other parameters constant. It was found that the expansion force and tubular surplus increase while the tubular thickness keeps thinning as the expansion ratio increases ratio irrespective to the mandrel angle and the friction coefficient. However, the tubular length variation shows similar behavior as that of the drag effect. It is important to note that the surplus deformation depends on the hardening behavior of tubular material. The difference in tubular final and desired inner diameters in terms of desired inner diameter is called surplus deformation. As a result of tubular expansion, the final tubular inner diameter will not be the exact required value. This happens due to the material's elastic recovery once the mandrel passes the expansion zone and the system dynamics of coupled system under consideration. The elastic recovery will be more in materials with lower hardening coefficient and hence influence the surplus deformation. This means that the hardening behavior of tubular material will have an effect on the magnitude of surplus deformation.

6. Anticipated failures and envisaged solutions

Occasionally, some of the mandrel geometric configurations combined with the irregular mandrel/tubular interaction as well large expansion ration may cause tubular failure. When a mandrel angle is 10°, the surface contact area is maximum between the mandrel and the tubular. If in such case the friction coefficient is high ($\mu = 0.4$) and expansion ratio approaches 35%, simulation results showed that the tubular will fail due to excessive wall thinning ultimately leading to rupture. Similar behaviors are also observed under other severe conditions. **Figure 18** shows two cases where the tubular has failed due to large magnitudes of expansion force, which subjects the tubular to extreme tension in post expansion zone. In order to avoid such failures, three solutions are proposed:

(a) Expand the tubular under compression instead of tension for the same mandrel angle, 10 and 20° (fix the top part of the tubular while running the mandrel from bottom upwards).

(b) Expand the tubular under plane strain condition for the same mandrel angle, 10 and 20° (fix the top and bottom parts of the tubular while running the mandrel bottom upwards).

(c) Use a spherical mandrel, if desire is for bottom-up tubular expansion under tension.

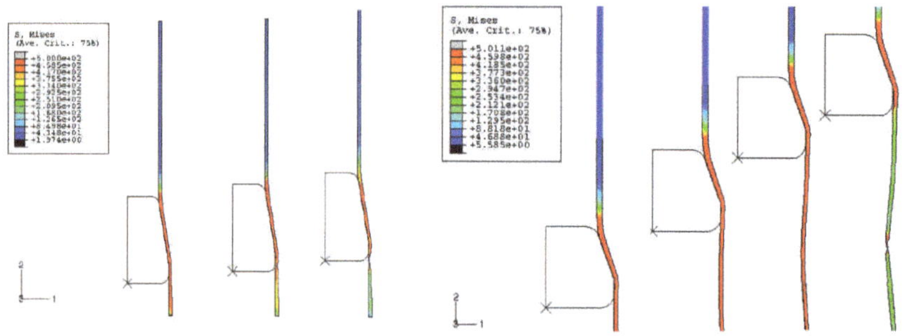

a) Mandrel angle of 10° (b) Mandrel angle of 20°

Figure 18. Simulation results showing tubular failure for $\mu = 0.4$ and 35% expansion ratio.

7. Rotating mandrel

Simulations have also been carried out to investigate the effect of the mandrel rotation on estimated drawing force required for expansion, surplus deformation, and thickness and length variations as shown in **Figure 19**. The results are obtained for 20° mandrel cone angle, 15% expansion ratio, $\mu = 0.1$ and 5 mm thick seal wrapped around the tubular. As can be observed from **Figure 19(a)**, the drawing force required for expansion decreases considerably due to the rotating mandrel. This could be due to the screw effect of the rotating mandrel. However, one must keep in mind that proper lubrication and additional power is needed to rotate the mandrel. The lesser magnitude of extrusion force for a rotating mandrel can be

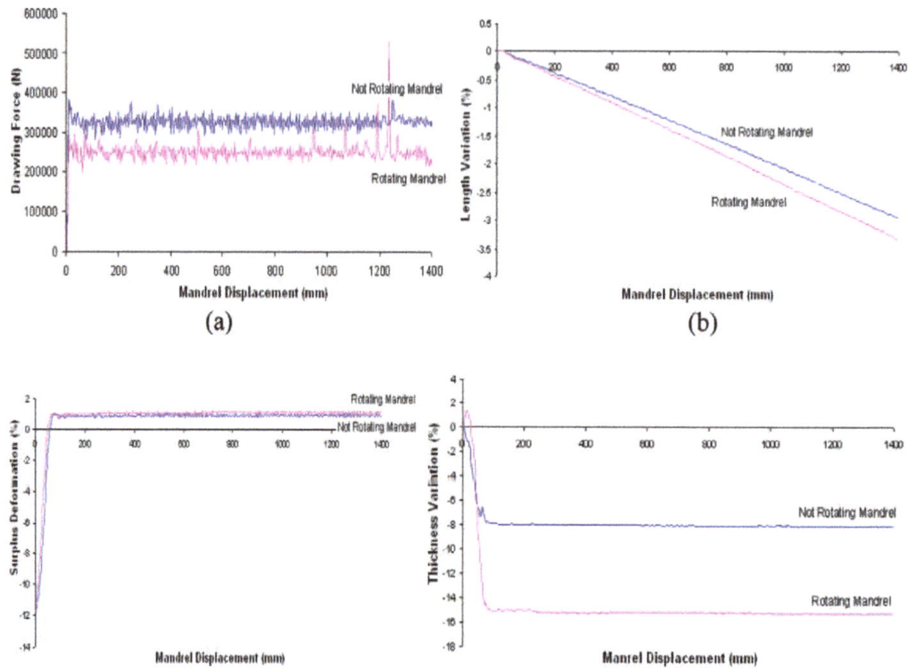

(a) (b)

(c) (d)

Figure 19. Plot of (a) drawing force, (b) length variation, (c) surplus deformation, (d) thickness variation versus mandrel displacement along the tubular for rotating and not rotating mandrel.

counter balanced by the extra power required for rotation. One has to estimate the total power requirement for the system to have a justifiable comparison between the rotating and nonrotating mandrel. Contrary to this, the thickness variation for rotating mandrel is approximately 47% higher than that of nonrotating mandrel (**Figure 19(d)**). A similar observation, but of lesser magnitude, is observed for length variation (**Figure 19(b)**). There is negligible effect on surplus expansion (**Figure 19(c)**). As pointed out earlier, higher magnitudes of variations in length and especially in thickness may cause structural instability in expanded tubular during its operational life.

Author details

Tasneem Pervez*, Sayyad Z. Qamar, Omar S.A. Al-Abri and Rashid Khan

*Address all correspondence to: tasneem@squ.edu.om

Mechanical and Industrial Engineering Department, College of Engineering, Sultan Qaboos University, Oman

References

[1] Koc M, Altan T. Prediction of forming limits and parameters in the tube hydroforming process. International Journal of Machine Tools & Manufacture. 2002;**42**:123-138

[2] Almeida BPP, Alves ML, Rosa PAR, Brito AG, Martins PAF. Expansion and reduction of thin-walled tubes using a die: Experimental and theoretical investigation. International Journal of Machine Tools & Manufacture. 2006;**46**:1643-1652

[3] Palengat M, Chagnon G, Favier D, Louche H, Linardon C, Plaideau C. Cold drawing of 316L stainless steel thin-walled tubes: Experiments and finite element analysis. International Journal of Mechanical Sciences. 2013;**70**:69-78. DOI: 10.1016/j.ijmecsci.2013.02.003

[4] Yang J, Luo M, Hua Y, Lu G. Energy absorption of expansion tubes using a conical–cylindrical die: Experiments and numerical simulation. International Journal of Mechanical Sciences. 2010;**52**(5):716-725. DOI: 10.1016/j.ijmecsci.2009.11.015

[5] Yan J, Yao S, Xu P, Peng Y, Shao H, Zhao S. Theoretical prediction and numerical studies of expanding circular tubes as energy absorbers. International Journal of Mechanical Sciences. 2016;**105**:206-214. DOI: 10.1016/j.ijmecsci.2015.11.022

[6] Filippov A, Mack R, Cook L, York P, Ring L, McCoy T. Expandable tubular solutions. In: SPE Annual Technical Conference; 3-6 October 1999; Houston, Texas; SPE 56500

[7] Hill R. The Mathematical Theory of Plasticity. New York: Oxford University Press; 1970

[8] Pervez T, Seibi AC, Karrech A. Simulation of solid tubular expansion in well drilling using FEM. Journal of Petroleum Science and Technology. 2005;**23**:775-794

[9] Pervez T, Seibi AC, Karrech A. Analytical solution for wave propagation due to pop-out phenomenon in solid expandable tubular system. Journal of Petroleum Science and Technology. 2006;**24**(8):923-942

[10] Pervez T, Qamar SZ, Al-Hiddabi SA, Al-Jahwari FK, Marketz F, Al-Houqani S, Velden M. Tubular expansion in irregularly shaped boreholes: Computer simulation and field measurement. Petroleum Science & Technology. 2011;**29**:735-744

[11] Marciniak Z, Duncan JL. Mechanics of Sheet Metal Forming. New York: Edward Arnold; 1992

[12] Stewart RB, Marketz F, Lohbeck WCM, Fischer FD, Daves W, Rammerstorfer FG, Böhm HJ. Expandable wellbore tubular. In: SPE Technical Symposium; 16 October 1999; Dhahran, Saudi Arabia; SPE-60766

[13] Fischer FD, Rammerstorfer FG, Daxner T. Flaring—An analytical approach. International Journal of Mechanical Sciences. 2006;**48**:1246-1255

[14] Yeh FH. Study of flaring forming limit in the tube flaring process. Journal of Strain Analysis. 2007;**42**(5):315-342

[15] Al-Hiddabi SA, Seibi AC, Pervez T. Stress analysis of casings expansion/post-expansion: Theoretical approach. In: ASME 6th Biennial Conference Engineering Systems Design and Analysis (ESDA2002/APM-92); 8-12 July 2002; Istanbul, Turkey

[16] Ruan CG, Maurer WC. Analytical model for casing expansion. In: SPE/IADC Drilling Conference; 23-25 February 2005; Amsterdam, Netherlands; SPE-92281

[17] Karrech A, Seibi A. Analytical model for the expansion of tubes under tension. Journal of Materials Processing Technology. 2010;**210**(2):356-362. DOI: 10.1016/j.jmatprotec.2009.09.024

[18] Al-Abri OS, Pervez T. Structural behavior of solid expandable tubular undergoes radial expansion process: Analytical, numerical and experimental approaches. International Journal of Solids and Structure. 2013;**50**(19):2980-2994

[19] DeLange R, Gandikota R, Osburn S. A major advancement in expandable connections performance, enabling reliable gastight expandable connections. SPE Drilling & Completion. 2011;**26**(3):412-418. DOI: 10.2118/135462-PA

[20] Seibi AC, Al-Hiddabi SA, Pervez T. Structural behavior of solid tubular expansion under large plastic deformation. ASME Journal of Energy Resources & Technology. 2005;**127**(4):323-326

[21] Gelfgat MY, Basovich VS, Tikhonov VS. Drill string with aluminum pipes designs and practices. In: SPE/IADC Drilling Conference; 19-21 February 2003; Netherlands; SPE-79873

[22] Pervez T, Qamar SZ, Seibi AC, Al-Jahwari FK. Use of SET in cased and open holes: Comparison between aluminum and steel. Journal of Materials and Design. 2008;**29**(4):811-817

[23] Binggui X, Yanping Z, Hui W, Hongwei Y, Tao J. Application of numerical simulation in the solid expandable tubular repair for casing damaged wells. Petroleum Exploration and Development. 2009;**36**(5):651-657

[24] Al-Abri OS, Pervez T, Al-Hidaabi SA, Qamar SZ. Analytical model for stick-slip phenomenon in solid tubular expansion. Journal of Petroleum Science and Engineering. 2015;**125**: 218-233

[25] Owoeye O, Aihevba LO, , Hartmann RA, Ogoke VC. Optimization of well economics by application of expandable tubular technology. In: IADC/SPE Drilling Conference; 23-25 February 2000; New Orleans, Louisiana; SPE-59142-MS

[26] Gusevik R, Merritt R. Reaching deep reservoir targets using solid expandable tubulars. In: SPE Annual Conference and Exhibition; 29 September-2 October 2002; San Antonio, Texas; SPE 77612

[27] Tubbs D, Wallace J. Slimming the wellbore design enhances drilling economics in field development. In: SPE Annual Technical Conference and Exhibition; 24-27 September 2006; San Antonio, Texas; SPE 102929

[28] Matthew J. World's first expandable monobore well system: Development, implementation and significance. In: International Petroleum Technology Conference (IPTC); 4-6 December 2007; IPTC-11587

[29] Ruggier M, Scott B, Urselman R, Mossor H, Van Noort RHJ. Advances in expandable tubing – A case history. In: SPE/IADC Drilling Conference; 27 February-1 March 2001; Netherlands; SPE-67768

[30] Bargawi RA, Zhou S, Al-Umran MI, Aghnim W. Expandable tubular successfully scab off severe casing leaks. In: SPE/IADC Middle East Drilling Technology Conference and Exhibition; 12-14 September; Dubai, UAE; SPE-97357

[31] Carlos E, Dean B, Waddell K. Increasing solid expandable tubular technology reliability in a myriad of down-hole environment. In: SPE Latin American and Caribbean Petroleum Engineering Conference;27-30 April 2003; Port-of-Spain, Trinidad and Tobago; SPE-81094

[32] Jennings L. Dynamic formations rendered less problematic with solid expandable tubular. In: IADC/SPE Drilling Technology Conference and Exhibition; 25-27 August 2008; Jakarta, Indonesia; SPE-114678

[33] Marketz F, Welling RWF, Noort RV, Baaijens TN. Expandable tubular completions for carbonate reservoirs. SPE Drilling & Completion. 2007;**22**(1):01-07. DOI: 10.2118/88736-PA

[34] Bathe KJ. Finite Element Procedures in Engineering. New Jersey: Prentice Hall; 2007

8

Stiff Vacuum Extrusion for Agglomeration of Natural and Anthropogenic Materials in Metallurgy

Ivan Kurunov and Aitber Bizhanov

Abstract

Recently developed concept of stiff vacuum extrusion (SVE) agglomeration for iron and steel making helps to innovate the briquetting technology and make it competitive with sintering. The results of the R&D in this field show that extrusion attributes very specific properties to the agglomerated products—BREX (extrusion briquettes), which favors their wide utilization in metallurgy—among them are quick strengthening of the BREX, their high hot strength, very low self-cost, and maintenance values. A set of the successful projects have been realized in iron making, ferroalloy production, and direct-reduced iron (DRI) production.

Keywords: extrusion briquettes (BREX), stiff vacuum extrusion (SVE), agglomeration, recycling, blast furnace, ferroalloys

1. Introduction

Most of the world's major metallurgical plants have accumulated millions of tons of blast furnace (BF) sludge during production. These wastes are typically recycled as a charge component for production of sinter in quantities limited by the allowable zinc input for the blast furnace (partial replacement of iron ore concentrate with low-cost sludge makes this recycling economically attractive). However, the addition of sludge in the sintering charge has a negative effect on sinter stability and quality.

Briquetting, and the subsequent use of briquettes in the blast furnace, offers a more promising and environmentally friendly way of recycling this sludge. In addition to improving sinter stability and quality, the carbon in blast furnace sludge in the briquettes is used as the reducing agent, whereas in the sinter process, it almost does not work. By briquetting sludge is removed from the sintering charge and directed into the blast furnaces to the extent that the zinc inputs to the blast furnaces do not exceed the established limits. Economic efficiency comes from replacing more expensive merchant ore pellets, reducing coke consumption by smelting the carbon-containing briquettes, and improving sinter quality after sludge is removed from the sinter plant's charge. Briquetting was widely used for agglomerating iron ore fines and waste in the 1920s. At one point, briquettes constituted between 30 and 40% of the charge for the West blast furnace plant in the Calbe (Germany). Briquettes also constituted 100% of the charge for a low-shaft blast furnace plant in Maxhütte, Germany. These briquettes were made from iron ore fines, coke, and limestone dust [1]. However, the advent of high-productivity iron ore and concentrates sintering method meant that briquetting was no longer competitively priced, due to the low capacity of the briquetting equipment.

Today, agglomeration of anthropogenic and natural metal-containing substances by cold briquetting, using mineral or organic binders, is becoming more prevalent. There are three basic technologies—roller pressing, vibropressing, and stiff extrusion with vacuum. Stiff extrusion with vacuum entails forcing a homogeneous wet mix (typical moisture content ranging from 12 to 18% of bulk material under 2.5–4.5 Mpa pressure through die holes.

Extrusion for agglomeration of ore and metallurgical wastes first occurred in the 1990s, when Bethlehem Steel commissioned a stiff extrusion line for briquetting 20 tons per hour of sludge and flue dust [2]. These briquettes were melted in Bethlehem Steel blast furnaces. The line operated until 1996, when it ceased due to the closure of the plant. This milestone was not thoroughly investigated by metallurgists until 2010, when the authors of the present paper began to study the characteristics of stiff extrusion and their influence on the metallurgical properties of extrusion briquettes (BREX) [3]. By April 2011, more iron-making specialists were evaluating industrial briquetting plant production, along with the use of stiff extrusion briquettes as a major component of the blast furnace charge.

Extrusion characteristics and the properties of briquettes are consolidated in the following comparative table (**Table 1**). Entries are based on the results of briquetting plants and data published online by the producers of briquetting equipment, including roller presses and vibropresses.

One can see that stiff vacuum extrusion (SVE) offers clear advantages over other briquetting technologies. Stiff vacuum extrusion requires significantly less cement binder than roller-pressed briquettes and is less dependent on the moisture content of the charge for briquetting. It also allows production of briquettes with a minimum size (diameter) comparable to the size of the sinter and pellets.

Characteristics of the process and properties of briquettes	Machines for briquetting and their characteristics		
	Vibropress	Roller press	Extruder
Maximum output (Mt/h)	30	50	100
Cement binder content (%)	8–10	15–16	4–8
Thermal processing of raw briquettes	80°C (16–20 h)	–	–
Waste generation	–	30% of production	–
Shape of briquette	Cylinder, prism	Pillow	Any
Dimensions (mm)	80 × 80	30 × 40 × 50	5–35
Moisture content of charge (%)	<5%	<10%	12–18%
Possibility of immediate stockpiling of raw briquettes	–	Possible	Possible

Table 1. Briquetting technology comparison.

2. Specific features of stiff vacuum extrusion (SVE) agglomeration

According to the classification adopted in the brick industry, "stiff extrusion" means the extrusion process, carried out at pressures of 2.5–4.5 Mpa and moisture contents of the formable mass going from 12 up to 18% (**Table 2**, Ref. [4]).

The process starts with the mixing of BREX constituents together with a plasticizing binder (bentonite) and water and further preparation of the mix for homogenization (souring). To

Type of extrusion	Low-pressure extrusion	Medium-pressure extrusion	High-pressure extrusion	
Designation used in structural ceramic industry	Soft extrusion	Semi-stiff extrusion	Stiff extrusion	High-pressure stiff extrusion
Extrusion moisture (%)	10–27	15–22	12–18	10–15
Extrusion pressure (MPa)	0.4–1.2	1.5–2.2	2.5–4.5	Up to 30
Penetrometer (N/mm^2)	<0.20	0.20–0.30	0.25–0.45	>0.30

Table 2. Types of extrusion and approximate values of the most important parameters.

achieve this process, all constituents and additives are removed from their bunkers or silos and loaded into feeding devices that precisely proportion each of them into a primary mixer where water is added. The mix is thoroughly mixed under high shear conditions. The next step is transportation of the moistened mix to the area of homogenization and curing of the mix for 24 h. After 24 h of souring, the material is loaded into Even Feeders, which feed the material in a regular way to the stiff vacuum extrusion briquetting line. This is followed by the secondary mixing of the moistened mix and binder, transportation of the prepared homogeneous mix to the extruder, and extrusion of the finally prepared mix to produce green BREX of a given strength.

The first step of the extrusion process is to blend cement binder and additional water with the soured mix. This is done via a secondary mixer that sits on top of extruder and discharges the mixed material into it (**Figure 1**). This mixer is designed to seal for vacuum as well.

The mixture enters the vacuum chamber and is partially agglomerated there due to the presence of high vacuum inside the chamber and removal of the air and moisture. The mixture is immediately crumbled, and its isolated particles fall down on blades of extruder auger. It is known [5] that the air adsorbed by surface of the particles of the plastic material in the form of multimolecular layers held by van der Waals bonding delays wetting them with water, even prevents compaction of mass, and contributes to the elastic deformation in plastic forming, contributing the lamination and microcracks detected during the drying and sintering of the extrusion products. Filling the pores, the air also prevents moisture penetration and separates particles. Vacuum removes the air from the pores and promotes closer contact of particles. The vacuum is maintained throughout the working volume up to the extruder dies. The vacuum level is at least 100 mm Hg (in absolute values). The combination of high mechanical values of auger pressure and the presence of vacuum, which deletes almost all compressible air from the material before it is subjected to pressure, leads to high values of the strength of the

Figure 1. General view of the extruder (bottom) and pug sealer.

green BREX, which allows their immediate transportation by conveyors and their stacking, practically without generation of any fines. In addition, it is known that under vacuum the viscosity of the cement paste decreased. This simplifies its uniform distribution and improves their interaction with water [6]. In combination with higher density of extruded mass, due to the removal of air, this leads to a decrease in consumption of cement binder.

BREX production ends with the exit of the elongated rods through the holes of the extruder die. These rods spontaneously break down during transportation by conveyors and further stockpiling into shorter fragments without generation of any dust (**Figure 2**). The decomposition of the elongated BREX during transportation by conveyors and discharge on the floor was described by Bizhanov et al. [7]. The reason for this is the bending of elongated BREX due to the gravity at the exit from the extruder that resulted in generation of 2–3 concentrators of the strain in the areas of future decomposition of this BREX.

Figure 2. Raw BREX at the exit of extruder and in the pile.

3. Investigation of BREX physical and metallurgical properties

We studied the process of strengthening of the BREX with combined binder (bentonite and cement) under natural conditions. For tests, we have chosen the BREX produced by industrial stiff vacuum extrusion briquetting line. The BREX contained (%) 47.2 of converter sludge, 28.3 of blast furnace dust, 18.9 of iron ore fines, 4.7 of Portland cement, and 0.9 of bentonite. Portland cement and bentonite were manually mixed in a dried state and added to a mixture before a pug mill with vacuum seal. For BREX samples, we daily measured compressive strength (on Tonipact 3000 (Germany) according to standard DIN 51067, open (apparent) porosity (vacuum method of liquid saturation according to standard DIN 51056), and density (on Mettler balance (the United States)). **Table 3** gives the results of daily measurements during 7 days. Apparent porosity and compressive strength values are given also in comparison in **Figure 3**.

A pronounced local peak is clearly visible in the BREX strength curve at the third day. The next day it changes into softening. During further storage, the strength increases. Note that, before softening, the BREX strength accounts for ~84% of the BREX strength after strengthening storage for 1 week. The change of open porosity almost follows the change in strength,

Strengthening time (days)	Apparent porosity (%)	Density (g/cm³)	Compressive strength (kgF/cm²)
1	31.5	2.42	24
2	25.4	2.66	45
3	32	2.43	63
4	27	2.44	52
5	27.2	2.45	56
6	26.2	2.45	57
7	26.8	2.46	59

Table 3. Change of physical properties of industrial BREX during strengthening.

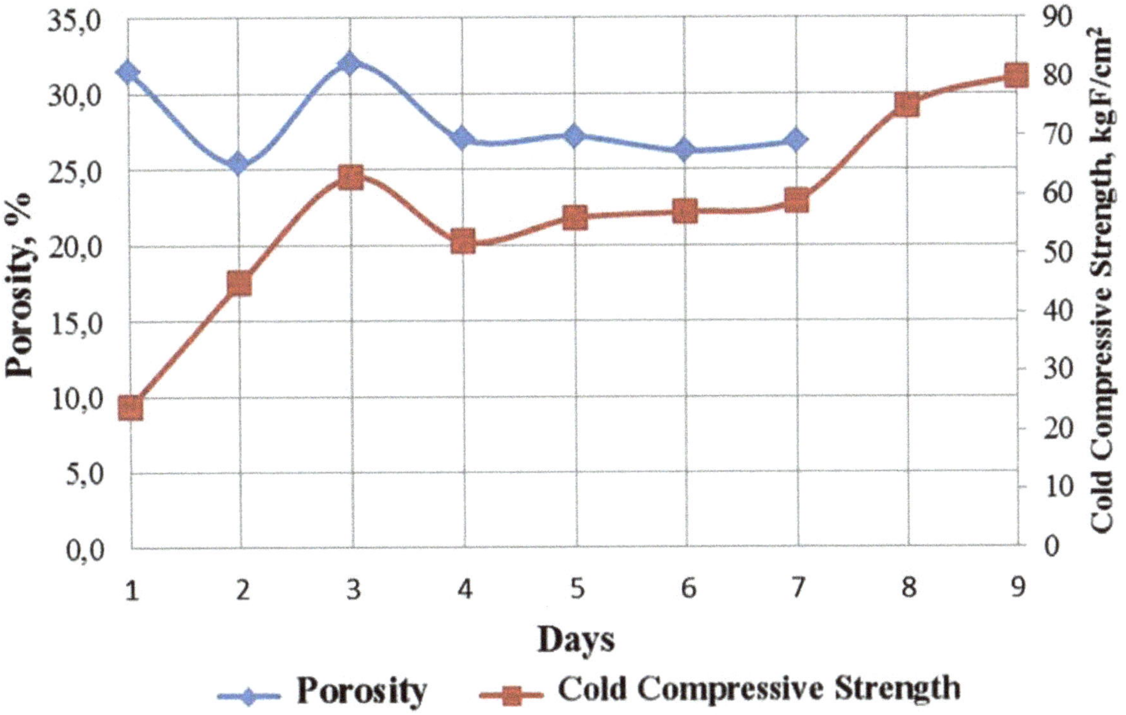

Figure 3. Change in strength and porosity of BREX during strengthening.

except for the first day of strengthening. The decrease in the porosity at that time is obviously related to the swelling of bentonite, which fills the pore space [8].

This behavior of BREX can be explained by the properties of a cement-bentonite binder and is related to the formation of coagulation structures in the cement-bentonite-water system, which leads to the modification of properties of the binder. Such structures are known to form in the gel-cement solutions used for the cementation of boreholes. The properties of the gel-cement solutions and the theoretical and practical aspects of formation and fracture of cement-bentonite systems are well known and systematized in, e.g., Ref. [9]. The driving force of the formation of such structures is the attraction of negatively charged bentonite particles to positively charged Portland cement particles, which results in their rapid coagulation and formation of a structure with suspended cement particles. Hydrated cement particles are gradually coated with an impermeable shell of flaky bentonite particles. The number of adsorbed bentonite particles is proportional to the activity of cement. During hydration, Portland cement particles grow in size, which leads to tension, break in the integrity of bentonite shells, and penetration of water into cement particles (i.e., to further hydration of cement and, apparently, adsorption of a larger amount of bentonite).

In order to identify possible impact of shear stress on the change of particle size distribution occurring in conditions of stiff vacuum extrusion, we have compared the results of pulverization of coke in three different ways: in a hammer mill, smooth roller crusher, and double shearing through the shearing plate of the extruder (**Figure 4**).

The degree of grinding of coke breeze appeared to be maximal after double shearing through a shearing plate in an extruder. The effect of deep grinding in this case is reached due to the application of high shear stresses. The use of a hammer mill for such a material was found to be ineffective, and the granulometric composition of the ground material differed weakly from that of the initial coke breeze.

Based on this we decided to compare the compressive strength values of the BREX produced from these three differently processed materials. Three sets of BREX were manufactured using laboratory extruder with the same composition: 94% of coke breeze, 5% of cement, and 1% of bentonite. The only difference is being how the coke breeze has been pulverized. Those were the BREX #1 (smooth-rolled coke breeze), BREX #2 (hammer-milled coke breeze), and BREX #3 (double-sheared coke breeze). The BREX samples were subjected to tensile splitting tests on a bench-type one-column electromechanical Instron 3345 tensile testing machine at a load of 5 kN. **Figure 5** shows the results of testing the specially prepared cylindrical specimens of BREX with 25 mm diameter and 20 mm height. It is seen that, for an approximately the same load-bearing capacity of BREX specimens, their reactions to an applied load are different. The difference in maximum loads can be due to defects in the specimens. However, the difference in the characters of behavior may have radically different causes. BREX #2 demonstrates ductile fracture, which is indicated by existence of a yield plateau, i.e., horizontal component of

Figure 4. Shearing through the shearing plate extruder; view of shearing plate.

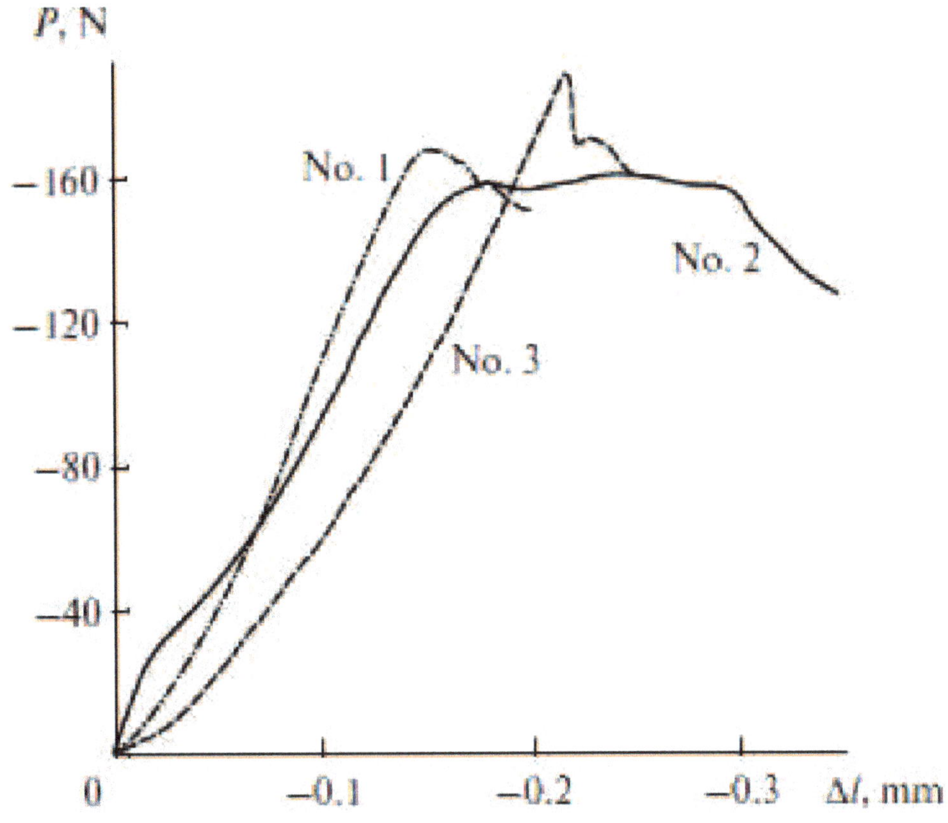

Figure 5. Results of the tensile splitting test of BREX; *P*, load; Δ*l*, displacement.

the curve of BREX #2. This phenomenon may be explained by a "relay race" transfer of gliding from grain to grain in terms of the Hall-Petch relation [10]. In this case, a grain boundary is a barrier to dislocation motion, which causes dislocation nucleation and development in a neighboring grain. In other words, the larger is the number of barriers to be overcome, the lower is the dislocation motion dynamics, and the higher is the crack development resistance.

One of the most compelling advantages of BREX with combined bentonite-cement binder is the possibility of their accelerated strengthening in comparison with the traditionally produced briquettes without addition of plasticizers. Composition of the BREX which we have subjected to three drops on the steel plate from 2 m height was as follows: A and B iron ore-sludge mixture (62.5%), mixture of flue dust and aspiration dust (30.5%), Portland cement (6%), and bentonite (1%); C- iron ore-sludge mixture (62.4%), mixture of flue dust and aspiration dust (30.1%), Portland cement (7%), and bentonite (0.5%); and D- iron ore-sludge mixture (62.2%), mixture of flue dust and aspiration dust (30.3%), Portland cement (6%), and bentonite (1.5%). Two modifications of BREX C were covered by tests using the addition of different amounts of water during mix preparation (50 and 100 ml). Components for all considered BREX were subjected to preliminary homogenization for 24 h in the form of hydrated mixture with the addition of bentonite. This method allows achieving a high degree of homogeneity of mixture properties for subsequent agglomeration. In some cases, the mechanical strength of BREX made of preliminary homogenized mix can significantly increase. Plasticizer's consumption can be considerably decreased during agglomeration of the preliminary homogenized substances.

Results of drop tests of the BREX after only 48 h of strengthening are presented in **Table 4**. It is known that the same strength level for the vibropressed briquettes on the cement binder can be achieved only after heat treatment at 80°C for 16–20 h or under the natural conditions (at a temperature of not less than 20°C) for 7 days [11].

For the assessment of the metallurgical properties of BREX, we have considered their low-temperature reduction disintegration indices (RDI). We have compared these values for two different types of BREX (**Table 5**)—blast furnace (BF) and basic oxygen furnace (BOF) sludge mixture with cement binder (BREX #1) and magnetite iron ore and coke breeze with cement and bentonite binders (BREX #2).

Testing of these two BREX types based on ISO 4696-1:2015 revealed a big advantage of BREX #2 compared with BREX #1 (**Table 6**). This can be attributed to the lack of any hematite phases in BREX #2 and to the presence of the secondary hematite in BREX #1. Crystal lattice of hematite is known to be subjected to the restructuring during reduction at low temperature causing mechanical stresses and disintegration of pieces of material that contain hematite. For comparison at the same time, we have measured following the same standard of the hot strength of sinters with basicity [$(CaO + MgO)/(Al_2O_3 + SiO_2)$] 1.2, 1.4, and 1.6. Hot strength of BREX #1 is comparable to the hot strength of sinters with basicity 1.2 and 1.4 (64 and 60%), which is due to practically the same content of secondary hematite. Hot strength of sinter with basicity 1.6 (77%) is larger than the hot strength of BREX #1 because of the presence in the sinter of this basicity of a new phase—calcium ferrites, which help to strengthen the sinter structure and to prevent its disintegration during low-temperature reduction. At the same time, the hot strength of BREX made of iron ore concentrate and coke breeze far exceeds the relevant indicators of all tested agglomerates.

For comparison we have studied the metallurgical properties of the hematite iron ore and coke breeze BREX (iron ore, 79%; coke breeze, 15%; cement, 5.55%; bentonite, 0.45%). Particles of this rich ore (Fe_{total}, 67.5%; SiO_2, 1.5%; $A1_2O_3$, 0.3%; CaO, 0.2%; MgO, 0.3%; S, 0.05%; P_2O_5, 0.05%) have a plate shape, which prevents their lump-forming capacity and makes the quality of the sinter lower. But this property of ore does not adversely impact on the quality of BREX and allows treating this ore as a promising raw material for the BREX

BREX parameters	A	B	C		D
Cement content (%)	6	6	7		6
Bentonite content (%)	1	1	0.5		1.5
Moisture content (%)	13.5	12.8	12.1	13.2	13.4
Drop strength testing (48 h) (%) of fines with size less than 5 mm	4.1	5.6	4.0	4.3	6.1

Table 4. Drop test results for the BREX after 48 h of strengthening.

BREX components	Mass share (%)	
	BREX #1	BREX #2
Portland cement	9.1	9.0
Coke breeze	–	13.5
Bentonite	–	0.9
BF sludge	54.5	–
BOF sludge	36.4	–
Iron ore concentrate	–	76.6

Table 5. BREX for RDI testing compositions.

Test material	RDI (%)
BREX #1 (1.93)	**61.9**
BREX #2 (basicity 0.75)	**96.5**
Sinter (basicity 1.2)	64
Sinter (basicity 1.4)	60
Sinter (basicity 1.6)	77

Table 6. Comparison of the RDI indices of BREX and sinter.

making. Mineralogical studies have shown that ore minerals are represented by hematite (Fe_2O_3) and rarely by splices of hematite with magnetite (Fe_3O_4). Silicates are most often observed in splices with iron minerals. To estimate the reduction process, polished sections of BREX samples were examined after their heating in a reducing atmosphere to temperatures: 900, 1100, and 1200°C. For the core of the BREX reduced by heating to a temperature of 900°C, the iron-containing phase is primarily represented by Wustite and magnetite (**Figure 6**, left), and the peripheral part contains linked metal iron particles with small inclusions of silicate phases (**Figure 6**, right).

In the peripheral part of the BREX heated to 1100°C, iron oxides have been fully reduced to the metal, and the metallic frame can be clearly visible (**Figure 7**).

Further heating to a temperature of 1200°C completes the process of BREX reduction of iron in its entire volume. In the core of BREX, iron is represented by metal and only partially in the form of Wustite. A small amount of inclusions of the unreacted particles of coke breeze (**Figure 8**) testifies to its abundance in the mix for BREX production even for the rich iron ore (67.5% Fe).

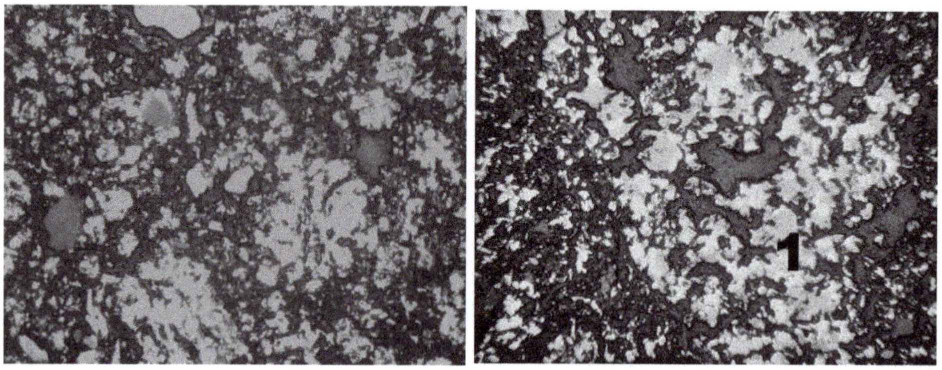

Figure 6. At left: microstructure of the core of the BREX reduced at a temperature of 900°C; reflected light, magnification × 100. At right: the formation of the metal frame by hematite grains (1) on the periphery of BREX (T = 900°C), the reflected light, magnification × 200, light gray-separate plots of the reduced Wustite and magnetite, and gray silicate phase.

Figure 7. Microstructure of the periphery of the reduced BREX at a temperature of 1100°C: white, metal; gray, transformed mineral phases of the cement binder; reflected light, magnification × 200.

Thus, with increasing temperature above 900–1000°C, a major role in the reduction of the iron oxides in the body of the BREX is played by carbon of the coke breeze; in the peripheral part of the BREX the metallic frame is being developed at this stage resulting from the oxide reduction by gas.. The presence of coke breeze particle in BREX after its heating in reducing atmosphere until 1200°C leads to the conclusion that it is necessary to maintain the carbon content of BREX in accordance with the stoichiometric ratio of C/O equal to or slightly greater than 0.3–0.5 relative to atomic oxygen content in iron oxides of BREX after their reduction to Wustite [12].

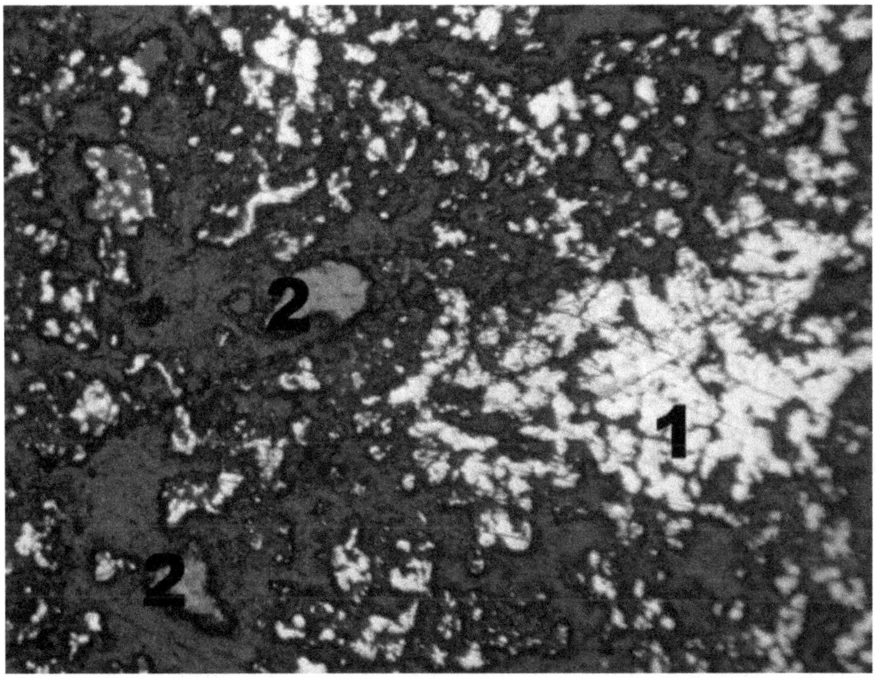

Figure 8. Microstructure of the core of the reduced BREX at a temperature of 1200°C. (1) metal, (2) coke breeze, gray, transformed mineral phases of the cement binder; reflected light, magnification × 200.

4. Application of stiff vacuum extrusion for the production of blast furnace charge components

Based on these results, the decision was made to achieve the maximum possible share of BREX in the charge of an industrial small-scale blast furnace with a volume of 45 m^3 (working volume 40.01 m^3) [13]. The blast furnace is equipped with a skip hoist with a skip volume of 0.5 m^3, double-cone charging device, hydraulic equipment for notch service, hot blast stoves, and a two-stage dry gas-cleaning system (dust collector and seven modules of bag filters). The blast furnace has eight air nozzles and one iron notch. External watering cools the furnace. Produced cast iron is immediately poured by the casting machine, and slag is granulated. Cast iron and granulated slag are shipped to customers by truck.

With two shifts working, the extrusion line produces 200 tons of BREX per day (12 castings per day). A required amount of the so-called washing BREX, made of manganese ore fines (−3 mm size) with 5% of Portland cement for binder, is also produced with stiff extrusion according to the operation schedule. It is known that in some cases, technologists were faced with "cluttering" of the hearth of the furnace due to the deterioration of coke's ability to filter when filling the voids between its particles by the pieces of slowly moving smelting products. It can result in combustion of air tuyeres, decrease of hearth heating, and other phenomena, which reduce the melting performance. One of the most effective means of combating this phenomenon is "flushing" the hearth by liquid slag containing FeO or MnO. Typically, the special sinter made with the use of mill scale is used as the "washing" material. Manganese briquettes can also be used for such purposes. The blast furnace melts 100–135 tons of BREX a day.

Operations began in May 2011 with BREX forming 10% of the charge, a percentage that gradually increased. When operating on a charge of 80% BREX and 20% iron ore, coke consumption decreased by 150 kg/t of cast iron (22%). This resulted from the decrease of carbon content in the BREX and withdrawal of the charge of raw fluxes. Lowering blast furnace performance by 15%, while switching to the new charge with BREX, was due to lower iron content in this charge (7.2%) compared to the charge consisting of iron ore and raw fluxes. Further increasing the share of BREX in the charge was not possible, due to an excessive increase in basicity of slags caused by the high basicity of BREX, which resulted from the presence of Linz-Donawitz sludge. During the development of the blast furnace regime with this new type of briquetted charge, we had to go to the new lower level of the furnace stockline because of difficulties with the dry gas-cleaning system. Gradually increasing the percentage of BREX in the blast furnace charge resulted in a lower temperature of furnace top gas and increased moisture content. As a result, bag filters become clogged with the wetted dust, and their regeneration by reverse pressure pulses had not reached the positive effect. Lowering the stockline helped increase the furnace top gas temperature, and the bag filter sticking stopped. Lowering the stockline had virtually no impact on the performance of blast furnace, primarily because iron oxides in the BREX are reduced by the dispersed carbon in the briquettes.

At the same time, the decision was made to increase the percentage of iron ore fine share in the BREX from 9.45 to 18.9%. This contributed to better blast furnace performance, due to better sintering and higher value of the activity of the "virgin" iron ore substance compared to the sludge and dust which have already undergone high-temperature processing. For more than two and a half years, the blast furnace worked successfully with a charge containing 80% of BREX. This blast furnace has worked with a 100% briquetted charge for the past 3 years.

Using BREX first as the primary and then as the only component of the blast furnace charge was possible because of their sound metallurgical properties and compatibility with the requirements of the iron-making process. This compatibility spans the life cycle of these briquettes, from the moment they exit the extruder die through the formation of cast iron in the blast furnace. BREX generate no fines moving from the extruder to the stockpile to the ready goods warehouse (loading these briquettes by forklift generates negligible amounts of fines). This eliminates the need for fines screening before charging, without any damage to the blast furnace. Skips with BREX do not contain any fines—pouring from the bunker to the skip and from the skip to the blast furnace-charging device of blast furnace generates no dust. Inside the blast furnace, BREX do not collapse when lowered from the top, preserving their integrity through softening and further melting in the cohesion zone. Results from various high-temperature tests, using BREX of different compositions in reducing atmospheres to investigate their mineralogical structure, confirm this conclusion [4]. All of the BREX kept their shapes when heated at a speed of 500°C/h up to 1150°C with a half-hour soak at that temperature and then cooled in an inert atmosphere.

The blast furnace operation parameters with different shares of BREX in the charge are given in **Table 7**.

Performance of blast furnace	100% iron ore	80% BREX	100% BREX
Consumption, kg/ton:			
Iron ore	1500	372	–
BREX	–	1425	1960
Limestone	150	–	–
Dolomite	144	–	29
Scrap	132	–	–
Quartzite	–	–	13
Mn ore BREX	–	19	75
Coke*	680	530	490
Fe_{total} in fluxed charge (%)	57.6	50.4	45.5
Capacity (ton/m^3 per day)	1.9	1.62	2.0
Blow temperature (°C)	925	900	1000
Blow pressure (kg/cm^2)	0.5	0.34–0.38	0.38–0.42
[Si] (%)	1.0–1.8	1.0–1.5	0.8–1.1
[Mn] (%)	0.2	0.4–0.5	0.7–0.8
[C] (%)	3.8–4.0	3.75–3.90	3.80–3.95
[S] (%)	0.050–0.060	0.038–0.050	0.038–0.042
Hot metal temperature (°C)	1380–1440	1400–1450	1410–1450
(CaO) (%)	34.86	33.12	38.0–39.0
(SiO_2) (%)	31.98	30.23	30.0–32.0
(Al_2O_3) (%)	23.87	17.98	16.0–18.8
(MgO) (%)	9.46	9.48	8.0–9.5
(FeO) (%)	1.01	1.26	0.6–1.15
(MnO) (%)	0.35	0.75	1.3–1

*It means the size of coke lumps 15-25 mm.

Table 7. Operation parameters of blast furnace with BREX in charge.

The data show that when working on a charge of 80% of BREX and 20% of ore, coke rate decreased by 150 kg/t of iron (22%) compared to furnace operation using 100% of iron ore. Coke rate reduction occurred because of carbon contained in the BREX, as well as the withdrawal of limestone and dolomite from the charge. Lower performance with a charge of 80% of BREX results primarily from decreasing iron content in this charge by 7.2% compared to the charge with the iron ore and raw fluxes. Working with a charge of 100% of BREX ultimately led to further reduction in the coke rate, due to the additional carbon of BREX, raising the blast temperature by 100°C. In addition, the use of the so-called washing manganese ore BREX reduced viscosity of slags and improved refinement of smelting products. As a result, furnace productivity increased by improving the structure of stock column, reducing primary slag viscosity and raising the overall pressure drop.

As can be seen, despite the high specific heat loss (due to the small size of the blast furnace and low blast temperature), the consumption of coke in the blast furnace at 100% of BREX does not exceed 500 kg/t, which corresponds to the modern high-efficiency furnaces with iron content of 58–59% in charge with a temperature of 1200°C and with the blast overpressure of gas on furnace top equal from 180 to 250 kPa. This is the consequence of the self-reducing nature of BREX (due to the presence of carbon in flue dust), as well as their basicity, which provides the necessary basicity of blast furnace slag. The latter allowed elimination of limestone use. Additions of quartzite and dolomite were applied to adjust slag basicity and magnesia content in the slag. As a result, coke consumption for the 100% of BREX charge fell by approximately 200 kg/t compared to the 100% of rich iron ore charge operation of the blast furnace. It can also be seen that Si content in pig iron has approached the level typical for large blast furnaces.

Five years of industrial operation, producing agglomerated products for a blast furnace, show that BREX obtained from natural and anthropogenic dispersed raw materials have optimal and adjustable dimensions, manageable chemical composition, and high metallurgical properties. Thus, these BREX can be seen as a new type of agglomerated and fluxed furnace charge component for blast furnaces. Unlike sinter and pellet production, stiff extrusion is environmentally friendly and completely waste-free, with neither gaseous nor solid emissions.

Results of investigation of the metallurgical properties of BREX, together with the unique experience of small blast furnaces using 100% of BREX in the charge, served as the basis for research for the feasibility and efficiency of stiff extrusion for large-scale modern blast furnaces. Research performed with mathematical modeling of the blast furnace used a simulation of the process and DOMNA software developed in the National University of Science and Technology MISiS. Blast furnace smelting was evaluated under the existing conditions of the industrial blast furnace of PJSC "NLMK" (volume 4297 m^3) [14].

Modeling was conducted for the equal composition of cast iron and its temperature ([Si] =0.4%; [C] = 4.8%; T = 1500 °C) and for equal reduction efficiency (degree of approximation of the composition of the gas to the equilibrium in the Wustite reduction zone). Results showed that carbon contained in BREX reduced the consumption of coke by smelting compared with the base option by 10%. When working with a pulverized coal injection (PCI) furnace with 160–284 kg/ton coke consumption, injecting natural gas at the rate of 125 m^3/ton coke consumption reaches 354 kg/ton (**Table 8**).

Results of mathematical modeling of a blast furnace, using a charge containing BREX made of iron ore concentrate and coal, have shown high efficiency of the partial (50%) substitution of the sinter production by BREX production, along with a 10% reduction of coke rate and 50% decrease in gas and dust emissions during sinter production. Calculations show that, in view of the increased basicity of sinter produced (after the substitution of 50% of sinter production by BREX making), CO_2 emissions at the sinter plant will decrease by 32%, dust emissions by 50%, and sulfurous gas emissions by 43%.

The performance of the furnace	Traditional charge	Option 1	Option 2
Sinter consumption B2 = 1.7 kg/ton	1109	–	–
Sinter consumption B2 = 3.0 kg/ton	–	557	575
Pellets consumption (kg/ton)	546	557	541
BREX consumption (kg/ton)	–	557	575
Iron ore consumption (kg/ton)	–	17	–
Fe content in charge (%)	58.2	57.45	57.15
Coke rate (kg/ton)	391	354	284
Natural gas consumption (nm^3/ton)	125	125	35
Pulverized coal consumption (kg/ton)	–	–	160
Blow rate (m^3/min)	7483	7568	7340
Blow temperature (°C)	1240	1240	1240
O_2 content in blow (%)	30.5	30.5	30.5
Blast humidity (g/m^3)	10	10	20
Top gas (m^3/ton)	1545	1540	1470
Top gas pressure (kPa)	240	240	240
CO content (%)	24.4	24.9	26.2
CO_2 content (%)	23.2	22.6	23.9
H_2 content (%)	9.7	9.9	8.2
Slag ratio (kg/ton)	318	314	323
Slag basicity (B2)	1.01	1.01	1.02
Capacity (ton/day)	12,465	12,624	12,708
Capacity (ton/m^2 day)	92.48	93.66	94.3
Reduction efficiency (%)	94.2	94.2	94.2

Table 8. Blast furnace smelting simulation results (traditional charge and charge with iron and carbon containing BREX (one third of charge).

5. Application of stiff vacuum extrusion for manganese ferroalloy production

An industrial extrusion line in Ragland, Alabama, produced 2000 tons of BREX to investigate the applicability of stiff extrusion for ferroalloy production. A 1400 tons of manganese ore fines were shipped from the Republic of Georgia (Chiatura), with 600 tons of baghouse dust shipped from a ferroalloy plant in West Virginia to make the BREX. The bulk raw material was blended and then transferred to the extrusion plant feed tanks. The blend was fed into a Steele 75ADC extruder, where it was mixed with water and 3–5% of cement. The extruder and vacuum mixer had a combined 338 KW capacity. The extruder was equipped with multi-hole extrusion die with round openings of 25 and 30 mm diameter. Typical production parameters were a production rate of 55 metric tons per hour, moisture content of the green BREX 10.5%, and vacuum level of 100 mm Hg absolute. **Figure 9** shows the production of the BREX and their immediate loading onto truck and further discharge within 5–10 min after production.

Next, the BREX were shipped to the smelter by barge and were discharged at a pier. Trucks provided transportation from the pier to an open-batch stockyard, followed by a conveyor directly to the furnace bunkers. In total, transportation of the BREX to the ferroalloy (2000 km) entailed 20 handling operations over 30 days. Total fines generated during these operations (less than 6 mm) did not exceed 10%.

To initiate a full-scale industrial trial, a stable 27MVA capacity and 85-Mt/day average production of industrial submerged electric arc furnace (SEAF) was selected to run with the specific average energy consumption at 4200 kWh/Mt, manganese recovery rate at 80%, and manganese content of dump slag at 12–14%.

To initiate a full-scale industrial trial, a stable 27MVA capacity and 85-tons/day average production of industrial submerged electric arc furnace (SEAF) was selected, using specific average energy consumption at 4200 kWh/Mt, manganese recovery rate at 80%, and manganese content of the dump slag at 12–14%. For the accuracy of comparison of results of the furnace operation, with and without BREX, the furnace operated for 1 month. A weeklong period of furnace operation, without BREX, immediately preceded the pilot operating period to serve as the reference. The results of the full-scale trial of the silicomanganese production with BREX in the charge of the industrial SEAF appear in **Table 9**. Their detailed description is given in Ref. [15].

The primary positive aspect of the full-scale campaign, using BREX in the charge for smelting of merchandise silicomanganese, is that the furnace operated in a smooth, stable regime. Furnace top operation was characterized by good gas permeability all over the surface, without any charge downslide. The current load was uniformly distributed among three electrodes.

Figure 9. Production (at left), loading (in the middle), and stockpiling of the green BREX for ferroalloy production.

Indicator		Pilot industrial operating period phase						
		Reference period	1	2	3	4	5	6
Actual metal production over a period of time (t)	ton	816.323	298.4	277.1	196.6	570.6	397.2	757.2
	b.t*	839.670	300.7	285.8	199.5	584.75	393.2	767.8
Actual furnace performance (%)		98.9	97.6	97.4	99.7	96.2	98.6	93.6
Power consumption (MW)		3.339.27	1.078.2	1.085.4	734.7	2.120.8	1.536.3	2.821.48
Specific power consumption (kW × h/b. Mt)		3.977	3.586	3.798	3.682	3.627	3.908	3.675
Ore-2	ton(29%Mn)/b.t	1.106	0.933	0.892	0.714	0.635	0.626	0.492
	ton(48%Mn)/b.t	0.668	0.563	0.539	0.431	0.383	0.378	0.297
Ore-1	ton(49.5%Mn)/b.t	0.565	0.505	0.482	0.509	0.480	0.524	0.484
	ton(48%Mn)/b.t	0.582	0.520	0.497	0.525	0.495	0.540	0.499
BREX	ton(31.37%Mn)/b.t	0	0.077	0.164	0.273	0.387	0.571	0.605
	ton(48%Mn)/b.t	0	0.050	0.107	0.178	0.252	0.373	0.395
The total consumption of raw manganese ore	ton/b.t	1.671	1.515	1.538	1.496	1.502	1.721	1.581
	ton(48%Mn)/b.t	1.250	1.133	1.143	1.134	1.130	1.291	1.191
Coke (ton/b.t)		0.432	0.341	0.420	0.400	0.381	0.418	0.404
Quartzite (ton/b.t)		0.419	0.499	0.524	0.465	0.529	0.456	0.475
Briquettes based on Ferrosilicon manganese screenings (ton/b.t)		0.158	0.082	0.103	0.092	0.120	0.110	0.094
Scrap (Mn content in scrap, %) (ton/b.t)		0.358 (23.3)	0.601 (29.9)	0.462 (33.0)	0.477 (35.3)	0.461 (32.0)	0.455 (25.8)	0.373 (38.6)
Electrode mass (ton/b.t)		0.034	0.032	0.030	0.035	0.028	0.033	0.028
Manganese extraction from the ore component (%)		80.1	79.8	80.7	80.7	83.6	79.1	79.9

Basis ton.

Table 9. Furnace operation parameters during reference and full-scale trial periods.

Electrodes were deeply submerged and stable. Melt tapping took place on schedule and chemical composition of metal, and slag showed no significant changes. Replacing a substantial part of the lumpy manganese ore with BREX based on ore fines and aspiration dust led to an overall improvement of technical and economic process indicators.

Specific energy consumption during the test period decreased significantly. In the reference period, consumption per basic ton of the alloy was 3977 kWh, with the share of BREX replacing ore in the charges equal to 40% and specific energy consumption decreased up to 3675 kWh per basic ton (–7.6%). Another positive result of the full-scale trial relates to the increased manganese extraction from the ore. At 29% of BREX in the ore part of the charge, the manganese extraction was 83.6% versus the average extraction of 80% in the reference period of the furnace operation without BREX in the charge. Decreased extraction in the period preceding the final phase was not associated with the presence of BREX in charge and was the result of furnace downtime and problems with the electrode.

It is also important to know the relationship of the BREX share in the ore part of the charge with the specific productivity of the furnace, expressed in the basis of ton per unit of electricity consumed. The best performance is achieved when percentage of briquettes in the ore part of the charge ranges between 30 and 40%. In general, results of the full-scale trials provide reason to consider BREX based on manganese ore fines and baghouse dust from gas cleaning as a viable charge component for silicomanganese smelting. Three more new stiff extrusion lines for briquetting were recently built for ferroalloy production.

Applicability of stiff extrusion for briquetting has also been studied for direct-reduced iron (DRI) production [16]. Results indicate that BREX could achieve a metallization degree comparable with metallization of traditional DRI. It has also been demonstrated that BREX may be considered as a charge component for rotary hearth furnaces (RHF) [17].

6. Conclusions

The main conclusions on the results of R&D of stiff vacuum extrusion applicability for metallurgy are as follows.

We have discovered the special nature of BREX strengthening curing due to the application of bentonite-cement binders, leading to the existence of a local peak of compressive strength after 48 h of a strengthening aging. The utilization of combined bentonite-cement binder promotes more rapid strength gain in comparison with traditional briquettes. We have found that stiff vacuum extrusion technology can be competitive as the technology of the fine iron ore containing material agglomeration, which is also able to serve as a partial or complete substitute for sinter production. Industrial experience of 3 years of continuous operation demonstrated the economically efficient operation of blast furnaces with 100% of BREX in their charge. Full-scale industrial trial confirmed the efficiency of BREX utilization as the charge components of submerged electric arc furnaces (up to 40% of the charge). Stiff vacuum extrusion is the only technology to agglomerate efficiently ore concentrates and aspiration dusts of ferroalloy production. This conclusion has been confirmed by the practical experience of the recently commissioned stiff vacuum extrusion briquetting plants.

Author details

Ivan Kurunov[1] and Aitber Bizhanov[2*]

*Address all correspondence to: abizhanov@jcsteele.com

1 Novolipetsk Steel (NLMK), Lipetsk, Russia

2 J. C. Steele & Sons, Inc., Statesville, NC, USA

References

[1] Ravich BM. Briquetting in Ferrous and Non-Ferrous Metallurgy. Moscow: Metallurgy; 1975. 232. (In Russian)

[2] Steele RB. Agglomeration of steel mill by-products via auger extrusion. In: Proceedings of 23rd Biennial Conf. IBA; Seattle. WA. USA; 1993. pp. 205-217

[3] BREX. Russian Certificate on the trademark (service mark) #. 498006, The Rightholder A. Bizhanov, priority 02.03.2012

[4] Frank H, editor. Extrusion in Ceramics. Berlin, Heidelberg, New York: Springer; 2007. 468 p

[5] Moroz II. Construction Technology of Ceramics. Moscow: Ekolit; 2011. 384. (In Russian)

[6] Rouzhinskyi S, et al. All About Foam Concrete. St. Petersburg: JSC "StroyBeton"; 2006. 630. (In Russian)

[7] Bizhanov AM, Kurunov IF, Durov NM, Nushtaev DS, Ryzhov SV. Mechanical strength of brex. Part II "Metallurg". Metallurgist. 2012;**10**:36-40

[8] Komine H, Ogata N. New equations for swelling characteristics of bentonite-based buffer materials. Canadian Geotechnical Journal. 2003;**40**:460-475. DOI: 10.1139/T02-115

[9] Jones GK. Chemistry and flow properties of bentonite grouts. In: Jones GK Symposium on Grouts and Drilling Muds in Engineering Practice. Butteworths. L.: Inst. Civil Engineers; 1963. London. pp. 22-28

[10] Malygin GA. Plasticity and strength of micro- and nanocrystalline materials. Phys Solid State. 2007;**49**:787-788

[11] Dorofeev GA, Barsukova EA. On the choice of rational agglomeration of anthropogenic and natural materials. Chernaya Metallurgiya. 2015;**12**:73-79

[12] Kurunov IF, Filatov SV, Bizhanov AM. Evaluation of efficiency of ore-coal brex in BF via mathematical modelling. Metallurg. 2016;**10**:23

[13] Bizhanov AM, et al. Blast furnace operation with 100% extruded briquettes charge. ISIJ International. 2015;**55**:175-182

[14] Kurunov IF, Yaschenko SB. Calculation Method of Technical and Economic Parameters of Blast Furnace. Scientific Works of the Moscow Institute of Steel and Alloys; Moscow. 1983. 152 p

[15] Bizhanov AM, Kurunov IF, Dashevskyi VY, Pavlov AV. Extruded briquettes – new charge component for the manganese ferroalloys production. ISIJ International. 2014;**54**: 2206-2214

[16] Bizhanov AM, Kurunov IF, Wakeel AK. Behavior of extrusion briquettes (BREX) in midrex reactors. Part 1. Metallurgist; 2015;**59**:283-289. (Russian Original Nos. 3-4. March–April. 2015). DOI: 10.1007/s11015-015-0098-1

[17] Bizhanov AM, et al. High temperature reduction of the stiff vacuum extrusion briquettes under the ITmk3 conditions. ISIJ International. 2014;**6**:1450-1452

9

Extruded Aquaculture Feed

Efren Delgado and Damian Reyes-Jaquez

Abstract

Agro-industrial by-products are processed materials that can have high protein content or other nutrients. The agro-industrial by-products are traditionally sold at low prices for animal feed consumption. These residues of the agro-industry have a high concentration of nutritional and bioactive compounds, which can be applied as fishmeal substitutes. In this chapter, it is shown how extrusion can be an alternative process for aquaculture feed production, increasing digestibility, and functional properties of the aquaculture feed, such as water stability and floatability. The thermal process during extrusion decreases the antinutritional factors present in legumes or other agro-industrial by-products, such as trypsin inhibitors and lectins. This chapter reviews research related to new protein sources that can potentially complement or substitute fishmeal for aquaculture feed. The use of bean (*Phaseolus vulgaris*) protein and cottonseed meal as a fishmeal substitute are shown, as well as the optimization of the extrusion process for aquaculture feed production. The incorporation of plant protein into the aquaculture production contributes to a more sustainable process. The effect of the extrusion parameters on the final product and quality are explained.

Keywords: cottonseed meal, legumes, functional properties, fishmeal, extrusion, aquaculture

1. Introduction

The Food and Agriculture Organization (FAO) considers that about 16% of the consumed animal protein comes from fish proteins [1]. With increasing population, demand for fish consumption will increase. Aquaculture is a good alternative for wild fish production. Aquaculture is a growing economic activity with estimated sales in the USA in 2013 of $1.3 billion dollars [2]. About 43% of the world's fish production has increased in farms and has been increasing in the past decade, especially in Asia and Africa [3]. Worldwide the aquaculture production grew in 2013 to 97.2 million tons (live weight), with a value of 157 billion US dollars. Asia is the major aquaculture producer in the world. Aquaculture production, such as that of finfish and crustaceans, requires a high amount of fishmeal [1]. Fishmeal represents about 60% of the cost of aquaculture feed and is also a limited resource. The world market consumes about 68% of fishmeal for aquaculture products, such as shrimp, trout, salmon, and other species [4], and it is expected to grow in the next decades. Several investigations have looked into the use of alternative protein sources that could either supplement or substitute for fishmeal. Agro-industrial by-products have been successfully used for animal feed [5], but can also be applied in aquaculture.

Soybean has been successful to a certain point as a substitute for fishmeal as a protein source for different aquaculture species. Reports show the use of soybean in feeding trout and shrimp. About 47% of world soybean production is GMO-soy [6]. Up to 69% soybean in the global market is genetically modified, while 85% of the soy produced in the USA is genetically modified. Consumers are also looking for nonGMOs for consumption. Different legumes represent an alternative source of protein usable for aquaculture feed. Agro-industrial by-products have high protein concentration and are sold at low cost for animal feed. Bean (*Phaseolus vulgaris*) is a common legume grown and consumed in many countries and different climates. Bean plants have low water requirements and are a staple food in many areas of the world. Small and damaged beans have no economic value and represent an agricultural by-product. Beans are an excellent protein source for aquaculture feed if processed thermally to inactivate the antinutritional factors present in the kernel [7, 8]. Cottonseed meal is a by-product of the oil industry. After oil extraction, the cottonseed meal (CSM) can have a protein content of up to 55%. CSM is sold at low prices for cattle feed and other small ruminants. The presence of gossypol, an antinutritional agent in CSM, limits its use in aquaculture. The breeding program has developed cotton varieties with low gossypol content, acceptable for use in aquaculture [9]. Some aquaculture species have low amylases activity, which limits the enzymatic breakdown of starches.

Extrusion applies high-temperature in short processing times. Extrusion is an alternative for feed production, increasing digestibility, and functional properties of the aquaculture feed, such as water stability and floatability. The thermal process during extrusion decreases the antinutritional factors present in legumes or other agro-industrial by-products, such as trypsin inhibitors and lectins [8]. The chapter reviews agro-industrial by-products than can substitute or complement fishmeal for aquaculture feed. The optimization of the extrusion process for aquaculture feed production is discussed.

2. Chemical composition and diet requirements in extruded aquaculture products

Agro-industrial by-products are processed materials, where some of the main compounds have been either extracted or are products that do not comply with certain quality requirements. The oil extraction industry produces by-products with high protein, while distillery by-products have less sugar after fermentation, but high protein concentrations. The bean (*P. vulgaris*) processing industry discards small kernels with no commercial application.

The agro-industrial by-products are traditionally sold at low prices for animal feed consumption. These residues of the agro-industry have a higher concentration of nutritional and bioactive compounds, which can be either used as supplements in functional foods or extracted and used for nutraceuticals. After oil extraction, soybean meal, canola meal, and flax seed meal contain high concentrations of proteins, minerals, and fiber. The brewing and distillery industry also offers by-products with high protein concentration. Cottonseed meal is obtained after oil extraction from the cottonseed. Glandless cottonseed has a low concentration of gossypol and is suitable for consumption by humans as well as livestock and aquaculture products. The levels of gossypol present in glandless cottonseed meal are not toxic for monogastric animals. The concentration of protein in glandless cottonseed meal (GCSM) can be up to 55%, which is more than 100% protein increased, compared to the cottonseed. GCSM has more protein than canola meal and is comparable to soybean meal (**Table 1**). Soybean seeds can have 40% of protein [10], but its content increases after oil extraction. Cottonseed meal has low starch content and a high mineral content, which makes it an excellent complement for aquaculture feed. Small and cracked beans do not have the required quality for consumer acceptance, but have a high protein and starch content (**Table 1**). Although the protein content is lower than other agro-industrial by-products presented in **Table 1**, the high starch content makes it more suitable for extrusion than by-products with low starch content. The extruded bean flour can be used for human and animal consumption, as well as for the aquaculture feed industry. The extrusion process totally inactivates the trypsin inhibitors and lectins in the bean flour [8].

Legumes have high protein content, and can potentially substitute for fishmeal. Bean flour can substitute for fishmeal in aquaculture feed. A balanced aquaculture feed should contain

Product	Fat (%)	Crude protein (%)	NFE (%)	Mineral content (%)	Crude fiber (%)
Soybean meal[4]	1.59	48.3	40.1	6.53	3.5
Canola meal[4]	2.8	36.1	43.3	6.3	11.5
Flax seed meal[4]	2.2	38.9	46.6	7.0	5.3
DDGS*[5]	10.9	30.2	53.1	5.8	8.8
Cotton seed[3]	20.71	23.1	31.8	4.85	19.54
Cottonseed meal[2]	12.6	41.0[3], 55.3	8.2	7.8	1.0
Bean (*P. vulgaris*)[6]	1.8	18.6	51.9**	3.7	11.0

[1]Solvent extracted.
[2]Cold pressed (Reyes-Jaquez et al. [11]), Gerzhova et al. [12].
[3]Gui et al. [13], Mujahid et al. [14].
[4]Khattab and Arntfield [15].
[5]Spiehs et al. [16].
[6]Delgado et al. [8].
*Total carbohydrates, Distiller's dried grains.
**Total starch content.

Table 1. Comparison of chemical composition of different agro-industrial by-products.

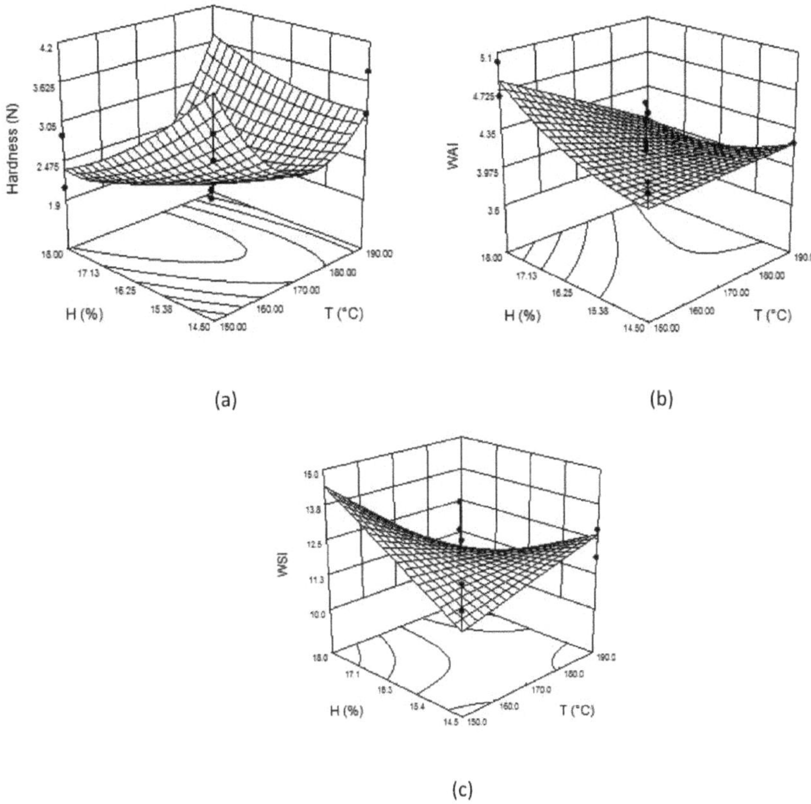

Figure 1. Effect of temperature and moisture content on the (a) hardness, (b) Water Absorption Index (WAI), and (c) Water Solubility Index (WSI) of extruded products from bean-nixtamalized corn flour. Design points above or below predicted value (reproduced with permission from Atienzo-Lazos et al. [7]).

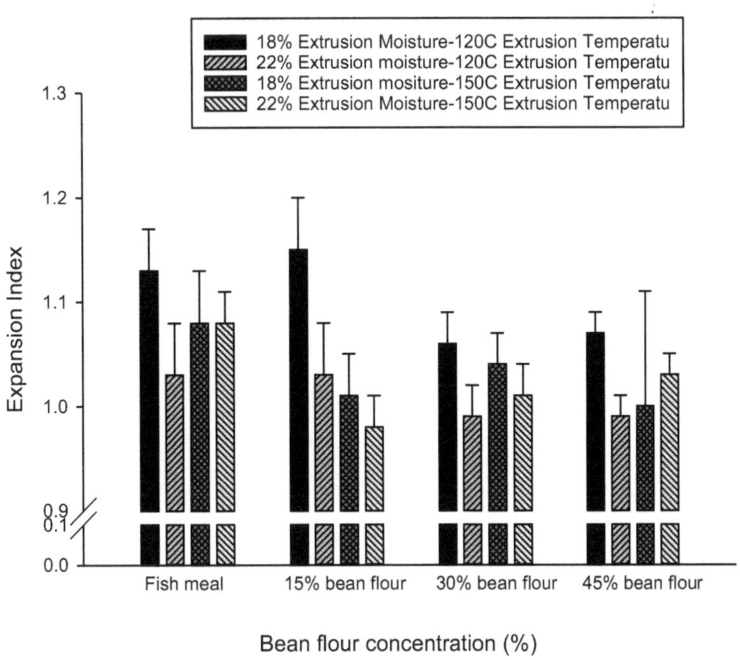

Figure 2. Expansion Index of extruded aquaculture feed at different extrusion moisture with fishmeal and different concentrations of bean flour (*P. vulgaris*). Vertical bars indicate standard deviation of the mean (Rodríguez-Miranda et al., 2014).

4. Extruded bean flour

Studies show that with a fishmeal substitution of 15, 30 and 45% bean flour or soy protein, there are no significant ($p > 0.05$) changes in protein content (**Table 3**). Even a 45% fishmeal substitution with bean flour/soy protein had similar protein content as the diet with no vegetable protein. The fat content of the aquaculture feed ranged between 15.9 and 18.8%, and the dry matter content was 91.6% or higher. The mineral content was significantly lower in the diets with 45% bean flour and with soy protein, compared to the diet with only fishmeal (**Table 3**). Bean flour is a better source of minerals for fish diet than soy protein concentrates. The functional properties of the extruded products are affected ($p < 0.05$) by the substitution of fishmeal with vegetable proteins. The Expansion Index (EI) decreases depending on the extrusion moisture. The EI decreases when substituting 30 and 45% of the fishmeal with bean flour at 18 and 22% extrusion moisture (**Figure 2**). When extruding at 22% moisture the EI and the bulk density (BD) were not affected ($p < 0.05$) by the substitution of fishmeal by bean flour (**Figure 3**) [18]. The Water Absorption Index (WAI) is also affected ($p < 0.05$) by the presence of bean flour (**Figure 4**). An increase in bean flour and decrease in fishmeal decrease ($p < 0.05$) the WAI and increase ($p < 0.05$) the Water Solubility Index (WSI) at 22% extrusion moisture. The WSI does not change ($p > 0.05$) at 18% extrusion moisture, and it even decreases at 22% extrusion moisture. The stability of the extrudates depends on the solubility and water absorption indices. The lower the WSI, the more stable the feed will be (**Figure 5**).

Another quality parameter in aquaculture feed is the sinking velocity of the product. The sinking velocity of the aquaculture feed is different for each aquaculture species. Some fish requires slow sinking feed that will resemble the movement of small fish or other living organisms. In the case of shrimp, the feed should sink to the bottom of the ponds for better use. Extrusion temperature affects ($p < 0.05$) the sinking velocity of the feed (**Figure 6**). Extrusion at 120°C increases ($p < 0.05$)

Extruded diet	Fat (%)	Crude protein (%)	NFE* (%)	Mineral content (%)	Dry matter (%)
With fish meal[1]	16.9[bc]	48.7[a]	29.2[a]	12.9[b]	91.7[b]
With 15% bean flour[1]	18.8[d]	41.9[a]	27.8[a]	11.4[b]	91.6[b]
With 30% bean flour[1]	16.7[ab]	40.9[a]	24.1[a]	10.4[b]	91.9[bc]
With 45% bean flour[1]	16.4[a]	38.7[a]	35.0[a]	9.7[a]	92.5[b]
With 15% soy protein[2]	17.7[c]	40.9[a]	32.4[a]	8.9[a]	92.6[a]
With 30% soy protein[2]	15.9[a]	42.2[a]	32.2[a]	9.5[a]	92.4[ac]
With 45% soy protein[2]	16.0[a]	45.3[a]	29.5[a]	91.9[a]	92.3[ac]

[1]Rodríguez-Miranda et al. [17, 18].

[a-c]Values with different letters in the same column are significantly different ($p < 0.05$).

*NFE = Nitrogen free extract.

Table 3. Chemical composition of extruded aquaculture feed with fishmeal and different concentrations of bean flour (Phaseolus vulgaris).

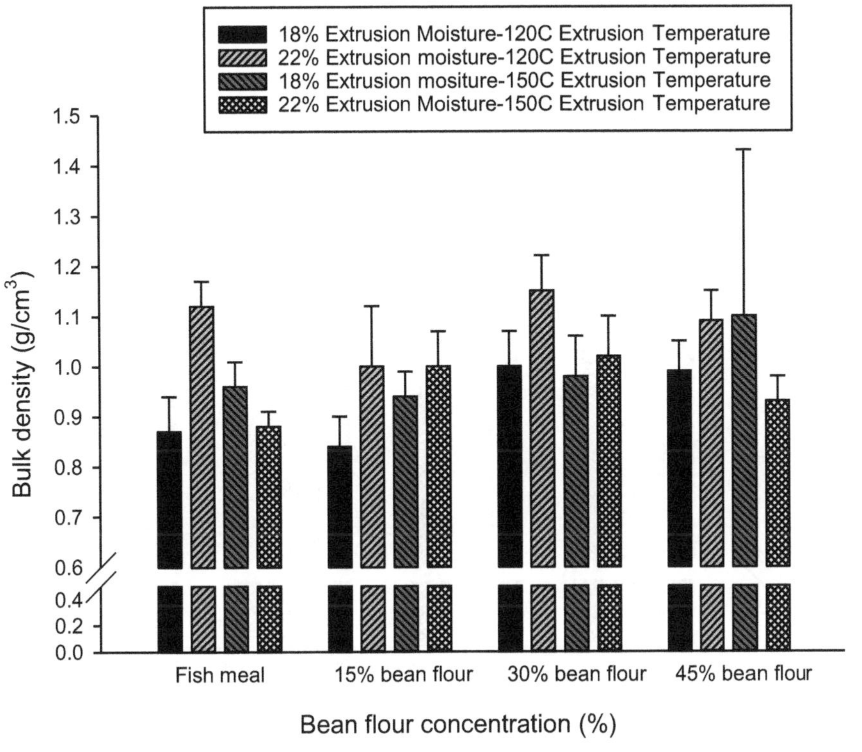

Figure 3. Bulk density of extruded aquaculture feed at different extrusion moisture with fishmeal and different concentrations of bean flour (*P. vulgaris*). Vertical bars indicate standard deviation of the mean (Rodríguez-Miranda et al., 2014).

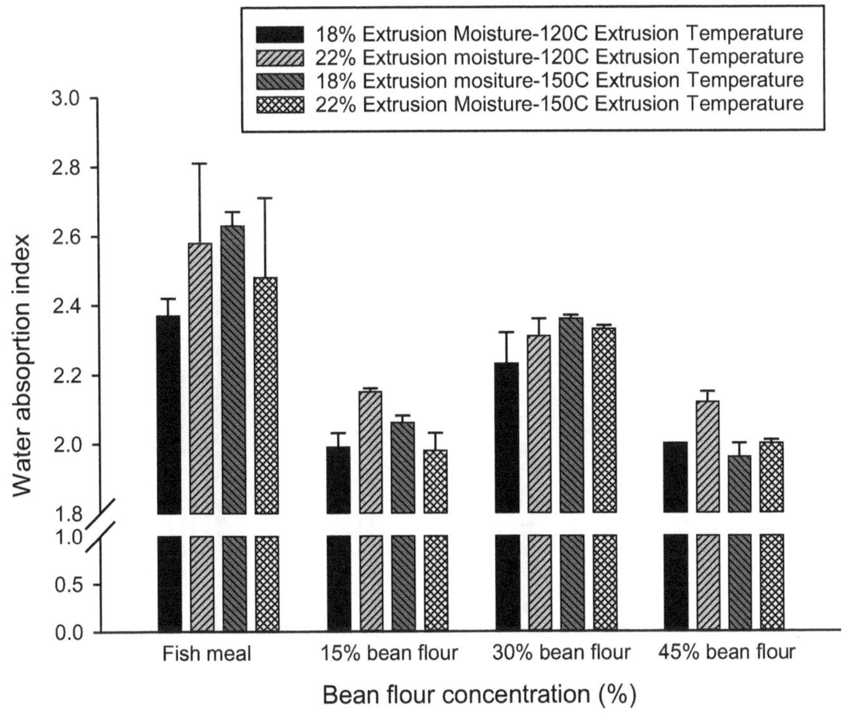

Figure 4. Water Absorption Index of extruded aquaculture feed at different extrusion moisture with fishmeal and different concentrations of bean flour (*P. vulgaris*). Vertical bars indicate standard deviation of the mean (Rodríguez-Miranda et al., 2014).

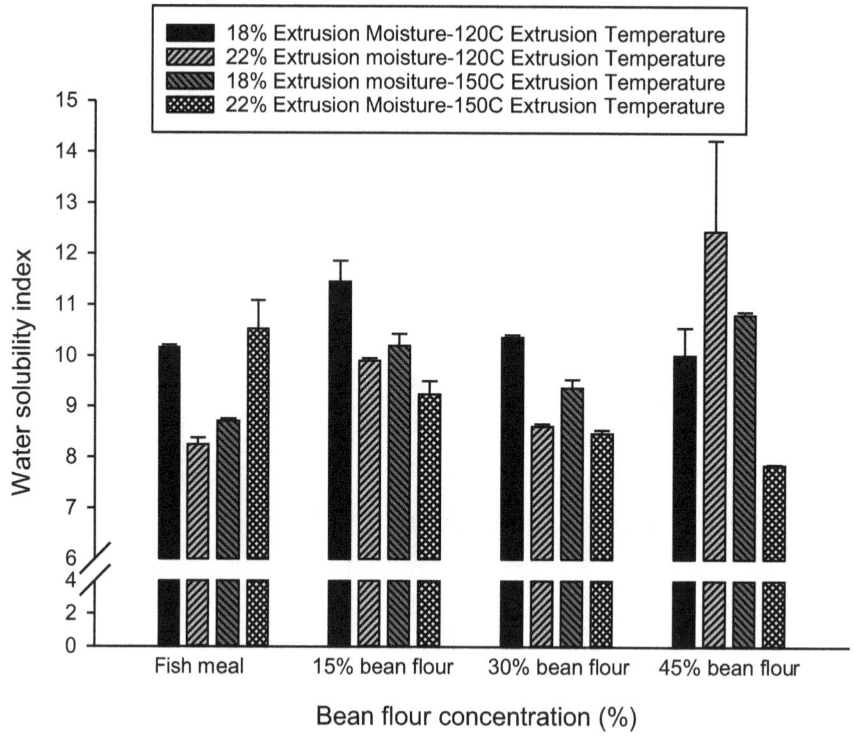

Figure 5. Water Solubility Index of extruded aquaculture feed at different extrusion moisture with fishmeal and different concentrations of bean flour (*P. vulgaris*). Vertical bars indicate standard deviation of the mean (Rodríguez-Miranda et al., 2014).

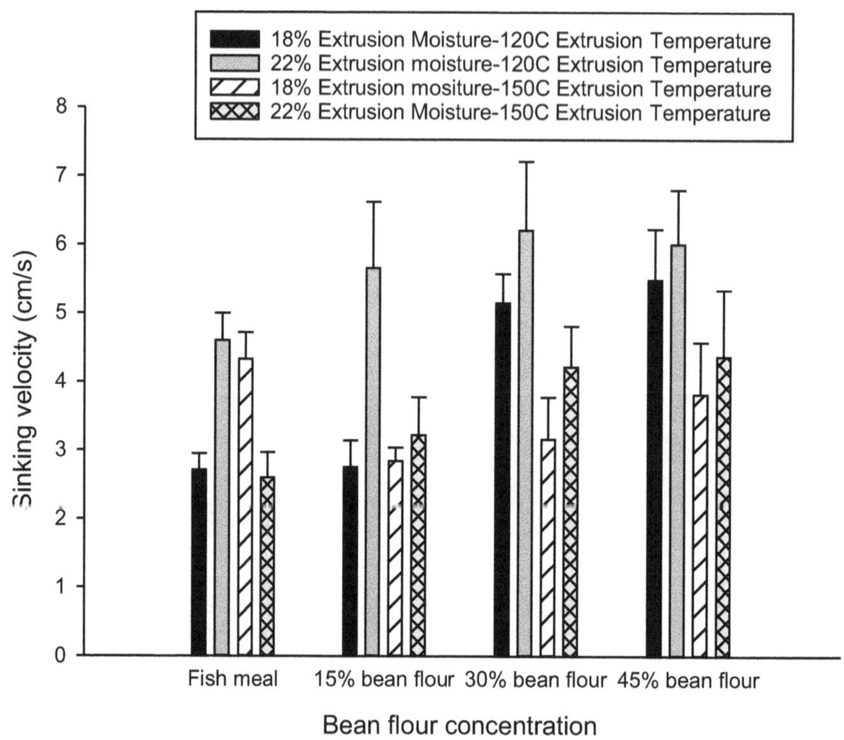

Figure 6. Sinking velocity of extruded aquaculture feed at different extrusion moisture with fishmeal and different concentrations of bean flour (*P. vulgaris*). Vertical bars indicate standard deviation of the mean (Rodríguez-Miranda et al., 2014).

the sinking velocity compared to extruded feed at 150°C. Bean flour can also affect ($p < 0.05$) the sinking velocity of the extruded feed. When feed contains bean flour and is required to have a low sinking velocity, the most recommended extrusion temperature is 150°C.

The independent variables to be considered in aquaculture extrusion are temperature, moisture content, and screw speed. Based on the independent variables and the dependent variables (Expansion Index, bulk density, and sinking velocity), in which the EI should be between 0.88 and 1.11, the bulk density ranges between 0.55 and 0.97 g/cm^3 and the sinking velocity is required to be between 2 and 6.2 cm/s. The best extrusion conditions with a single laboratory extruder (Brabender, Germany) are 120°C, 22% moisture content at a screw speed of 140 rpm with a diet formulation containing 62% fishmeal and no vegetable protein. Diets containing 15 and 30% bean flour require less moisture (18%), but the same temperature and screw speed, to obtain an optimum aquaculture feed. Diets containing 45% bean flour and 15% soy protein are best extruded at 120°C and 18% moisture at a lower screw speed (80 rpm). The lower mineral content appears to affect in this case the screw speed. A sharp increase in soy protein concentrate of 30 and 45% requires high extrusion temperatures of 135 and 150°C, respectively. Extrusion of 30% soy protein requires 20% moisture content and a screw speed of 110 rpm. The moisture requirements are also high (22%) as well as the screw speed (140 rpm) for extrudates with 45% soy protein concentrates.

The specific mechanical energy (SME) is the necessary energy in the form of work in the extrusion process. In aquaculture feed, the SME is affected ($p < 0.05$) by the extrusion temperature. High extrusion temperature decreases ($p < 0.05$) the SME. The extrusion moisture does not affect ($p > 0.05$) the SME, except with samples containing 15% bean flour. When extruding samples containing bean flour, the increase in screw speed will increase ($p < 0.05$) the SME, but not with those materials containing fish meal and no bean flour [19].

5. Starch pregelatinization

Pregelatinized and not pregelatinized starch has been added to balanced aquaculture feed to study the effect of pregelatinization on the functional properties of the extrudates. The studies have shown that pregelatinization of starch before extrusion has a positive effect ($p < 0.05$) on the Water Solubility Index (WSI) (data not shown). The WSI decreases and can be beneficial for the aquaculture industry. Pregelatinization does not affect ($p > 0.05$) the Water Absorption Index of the extruded feed, while the sinking velocity of the extrudates does not increase ($p > 0.05$) compared to the diets with no starch. Adding starch to the extrudates will lower ($p < 0.05$) the sinking velocity of the extrudates, as long as the starch is not pregelatinized. Pregelatinized extrudates are harder than extrudates containing starch that have not been pregelatinized before extrusion. The Expansion Index (EI) of extrudates containing pregelatinized starch decreased ($p < 0.05$) compared to extrudates with starch which was not gelatinized. The decrease in EI was observed with 20% of starch content, but there were no differences ($p > 0.05$) with extrudates containing more than 50% starch [20].

Nixtamalization can also be used to pregelatinize starch. Nixtamalization is a traditional thermal treatment used for corn products in North and Central America, where corn kernels are cooked with CaOH, resulting in a pregelatinized dough suitable for extrusion. **Figure 7b** shows pregelatinized corn kernels, where the center of the kernels appears to be enzymatically degraded during the steeping time of the process. **Figure 7a** shows a bean starch kernel before extrusion, while **Figure 7c** shows the structural matrix of extruded bean/corn flours. Again we observe the protein structure, but also partial gelatinization of the starch kernels. Different raw material influences the final product characteristics. Crystallinity represents the structure arrangement and is mostly related to the starch structure in the kernel. Although corn has a higher starch content (about 65.5%) [8], than bean flour (51.9%), bean flour shows a higher crystallinity than nixtamal (**Figure 8**). Extrusion decreases crystallinity because of the gelatinization of the starch kernels, but the temperature and moisture content during extrusion also affect the percentage of the crystallinity of the extruded product. The crystallinity of the extruded product is related to the retrogradation of the starch. Low extrusion temperatures produce the lowest crystallinity of the end-product probably because of the lower degree of gelatinization. High extrusion temperatures yield a low crystallinity because of starch dextrinization during extrusion and lower retrogradation.

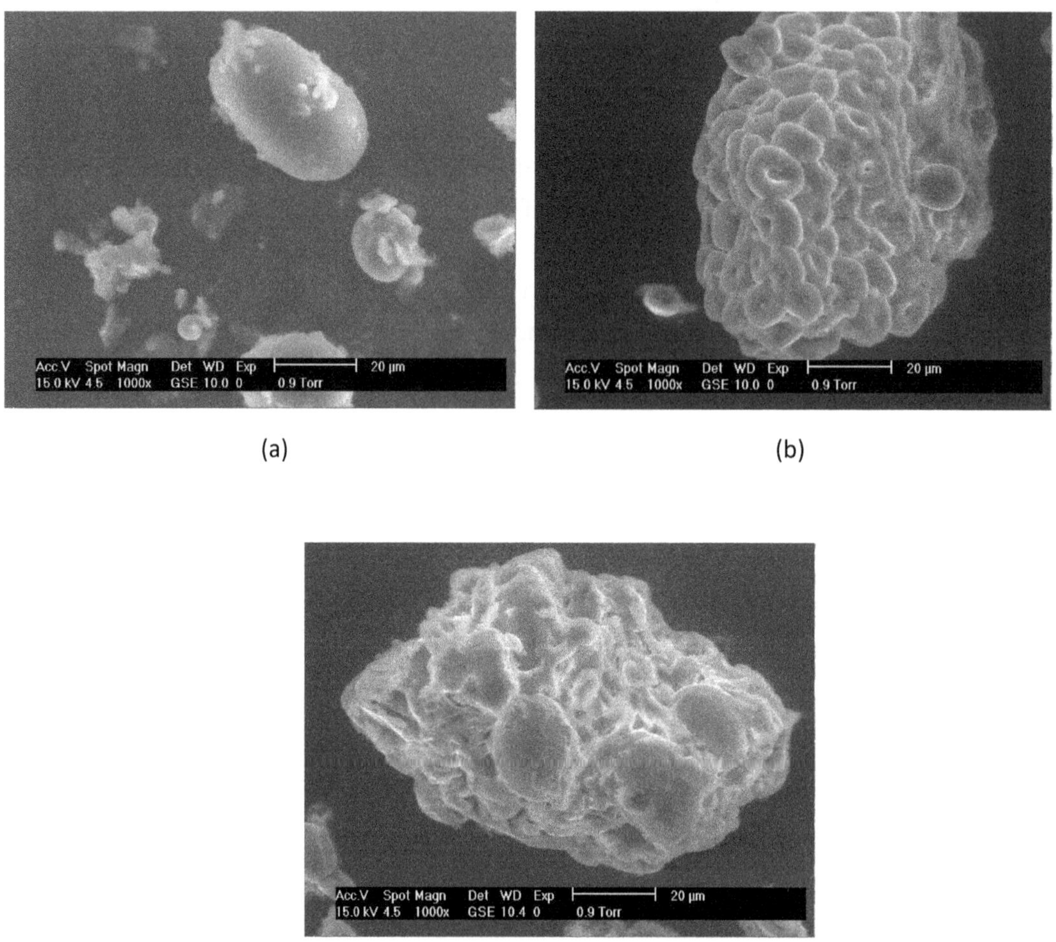

Figure 7. Scanning electron micrographs (1000 × magnification) of (a) bean flour, (b) nixtamalized corn flour, and (c) extruded bean-nixtamalized corn flour at 150°C and 14% H (C) (bar = 20 µm) (reproduced with permission from Atienzo-Lazos et al. [7]).

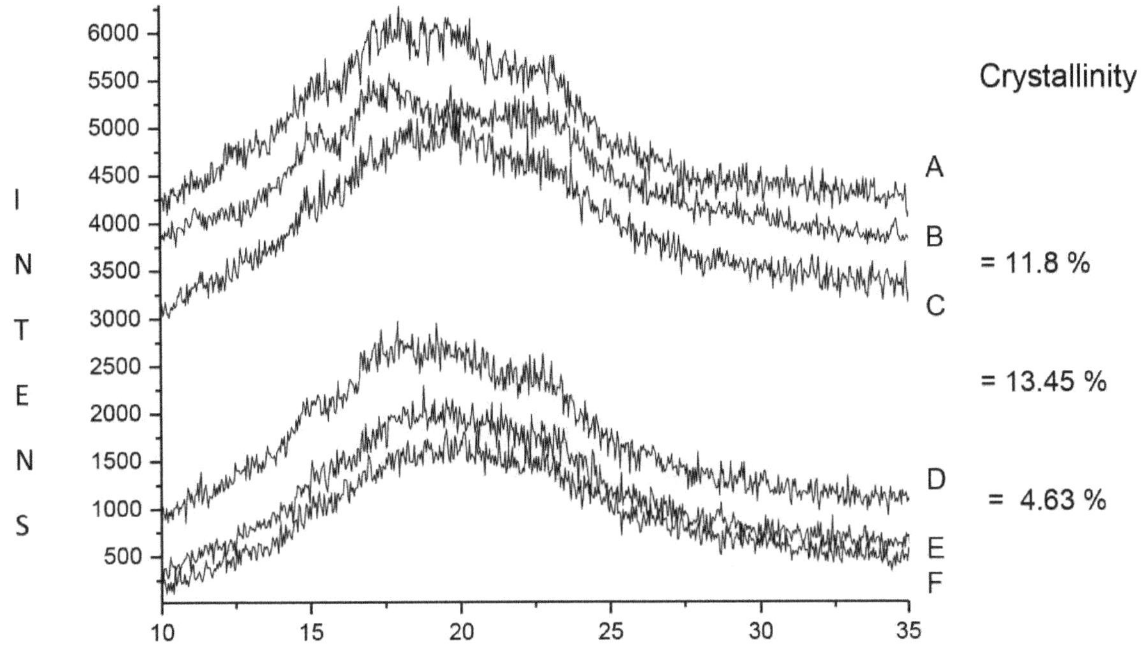

Figure 8. X-ray diffraction of A = Nixtamal flour, B = Bean flour, C = extruded flour at 150°C and 14.5% moisture, D = extruded flour at 170°C and 16.3 % moisture, E = extruded flour at 170°C and 18.7 % moisture, F = extruded flour at 190°C and 14.5% moisture (reproduced with permission from Atienzo-Lazos et al. [7]).

6. Feeding trials of extruded aquaculture feed

Studies of extruded trout feed show a final weight decrease ($p < 0.05$) after 32 days of feeding rainbow trout (*Onchorhynchus mykiss*) with bean flour. **Figure 9** shows a decrease ($p < 0.05$) in trout final weight and weight gained after being fed with extruded feed containing 15–45% of bean flour when compared to extruded feed-containing fishmeal. Trout fed with 45% of bean flour only gained 5.5% of weight in 32 days. Although weight gain is lower ($p < 0.05$) when fishmeal is substituted with bean flour, the survival rate was 100% for all the diets [21], indicating that bean flour can be used up to 30% as a fishmeal substitute. The feed conversion efficiency can be calculated as shown in Formula 1. The feed conversion efficiency (FCE) is highest ($p < 0.05$) with fishmeal (54.4%), and decreased ($p < 0.05$) to between 47.1 and 46.5% with 15 and 30% of bean flour, respectively. Extruded feed containing 15–30% bean flour shows an acceptable feed conversion efficiency. Feeds containing 45% of bean flour are not recommended for trout feed due to the low FCE Index.

$$\text{FCE} = (\text{Final weight(g)} - \text{Initial weight(g)}/\text{Ccomsumed feed(g)})*100 \qquad (1)$$

The condition factor or coefficient of condition *K* is a quality parameter of the fish and takes into consideration the weight and length. A condition factor above 1.6 shows that a fish has an excellent condition. The diets containing fishmeal and bean flour (15 and 30%) had a similar *K* factor (**Figure 10**). The *K* factor increased ($p < 0.05$) with 45% of bean flour. The diets containing fishmeal or 15 and 30% of bean flour had the same ($p > 0.05$) feed conversion ratio (FCR). The feed conversion ratio shows the inverse relationship between feed intake and weight gain of the fish and is related to the digestibility and metabolic use of the diet [22]. Trout fed with fishmeal or 15 and 30% of bean flour also had the same ($p > 0.05$) specific proteins and essential fatty acid, normally from fish oil, minerals, and vitamins. Aquaculture feed contains about 62% fishmeal, 20% wheat flour, 20% fish oil, 3.4% milk whey, 2.1% vitamins and minerals, and 0.5% choline chloride for cell function and structure (**Table 2**). The use of plant proteins should supply enough proteins to cover the nutritional requirements of the aquaculture products. Fishmeal is a limited resource, but alternative protein sources can substitute for fishmeal in aquaculture diets.

Composition	**Fish meal**	**Bean flour concentration**			**Soy protein concentrate**		
		15%	**30%**	**45%**	**15%**	**30%**	**45%**
Fishmeal	62.0	52.7	43.4	34.1	52.7	43.4	34.1
Bean flour	–	9.3	18.6	27.9	–	–	–
Soy protein concentrate	–	–	–	–	9.3	18.6	27.9
Wheat flour	20.0	20.0	20.0	20.0	20.0	20.0	20.0
Fish oil3	1.2	1.2	1.2	1.2	1.2	1.2	1.2
Dried whey	3.4	3.4	3.4	3.4	3.4	3.4	3.4
Choline chloride	5.0	5.0	5.0	5.0	5.0	5.0	5.0
Vitamin and minerals	2.0	2.0	2.0	2.0	2.0	2.0	2.0
Chromic oxide	1.0	1.0	1.0	1.0	1.0	1.0	1.0

*Rodríguez-Miranda et al. [17].

Table 2. Chemical composition of aquaculture diets (%)*.

3. Hardness and functional properties of extruded products

Hardness, Water Absorption Index, and Water Solubility Index are essential functional properties of aquaculture feed. The feed should have a certain hardness for the trout or shrimp to be able to eat it. The hardness of the extruded product depends on extrusion moisture and extrusion temperature. High extrusion temperature and high moisture content result in a hard extruded product. Softer products are the result of extruding at low temperatures and high moisture content (**Figure 1**). Aquaculture feed products need a specific Water Absorption Index (WAI) to facilitate consumption, while the Water Solubility Index (WSI) correlates well with the stability of the feed in an aqueous environment. The extruded products need to be stable in the water; a high WAI also produces a high WSI of the extrudates. Studies show that extruded products with lower WSI are obtained at low extrusion temperatures and low moisture content, which also produces harder extruded products. If we compare **Figures 1** and **2**, we can conclude that although crystallinity is lower in the product extruded at higher temperatures, the hardness tends to be high. The results indicate a probable high degree of denaturation, where proteins unfold, allowing protein to restructure into harder structure.

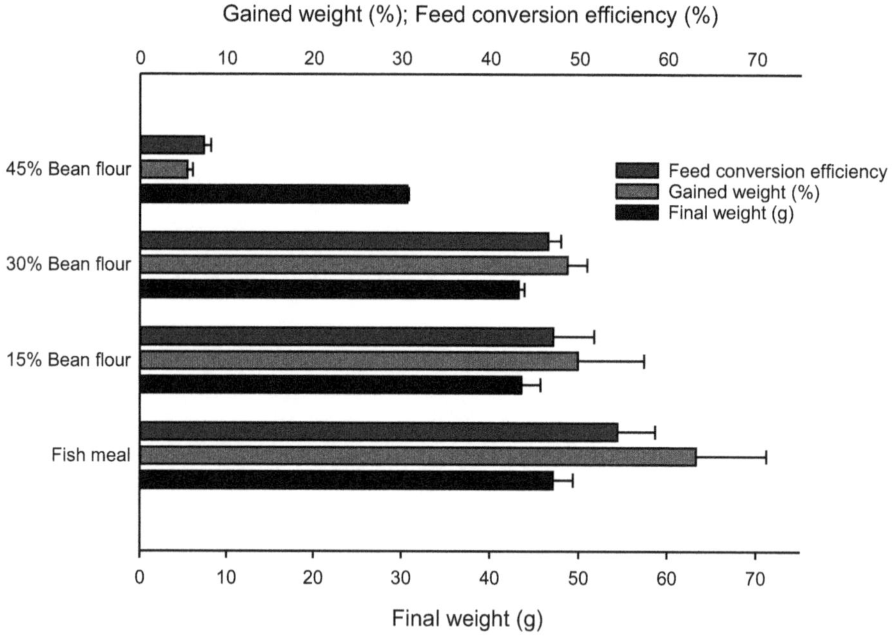

Figure 9. Feed conversion efficiency (FCE), gained weight, and final weight of rainbow trout (*O. mykiss*) fed with extruded aquaculture feed with fish meal or different concentrations of bean flour (*P. vulgaris*) for 32 days. Initial average weight = 29.1 g. Horizontal bars indicate standard deviation of the mean (adapted from Rodríguez-Miranda et al. [21]).

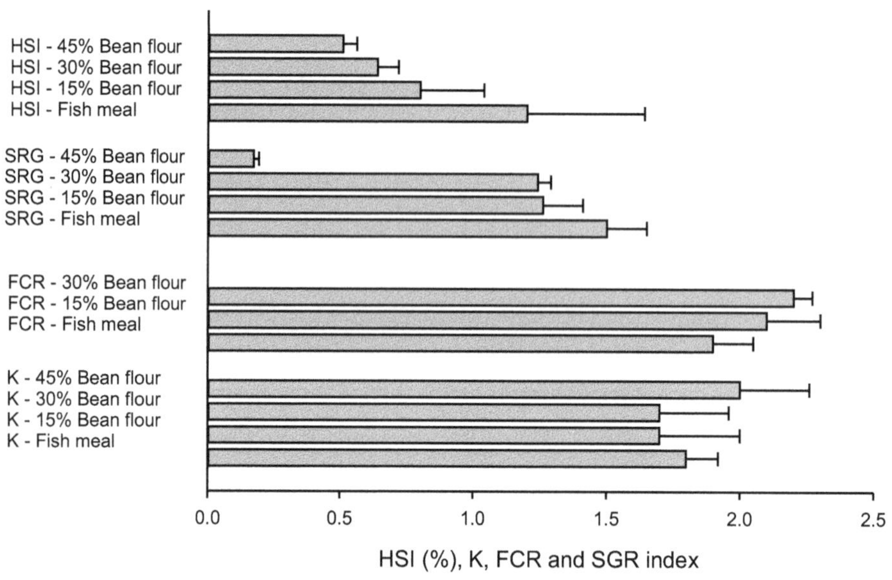

Figure 10. Condition factor (K), feed conversion ratio (FCR), Feed Conversion Efficiency (FCE) Index, and Hepatosomatic Index (HSI) of rainbow trout (*O. mykiss*) fed with extruded aquaculture feed with fish meal or different concentrations of bean flour (*P. vulgaris*) for 32 days. Initial average weight = 29.1 g. Horizontal bars indicate standard deviation of the mean (adapted from Rodríguez-Miranda et al. [21]).

growth rate (SGR), but not trout fed with 45% of bean flour in their diets. The Hepatosomatic Index (HSI) shows an indirect relationship between liver weight and body weight; it indicates the nutritional state of the trout. High HSI indicates a better fish nutritional condition. The fishmeal and the 15% bean flour diets have the highest HSI; there is no difference ($p > 0.05$) between the two diets. The diets with 30 and 45% of bean flour have lower ($p < 0.05$) HSI than the diets containing no fishmeal, and either no bean flour or 15% bean flour.

7. Color of extruded bean flour aquaculture feed

The L* values describe the lightness of the color of the sample on a scale of 0–100, where 0 = black and 100 = white. On the other hand, the a* values if positive (0–60) are related to a reddish color of the extruded product, if the a* values are negative (0–60), the feed tends to be greener. The closer the values are to zero the more the color tends to be neutral. The values with the highest ($p < 0.05$) luminosity have the lowest a* values. The samples with higher ($p < 0.05$) L* correspond to the extrudates with 45% of bean flour, probably because of the presence of more starch in the sample. These samples also have the lowest ($p < 0.05$) a*. The samples with the lowest L* are the samples without bean flour, except the sample extruded with 15% bean flour, 150°C, and 22% moisture content, which also has a low L*. The samples with less bean flour and more fishmeal tend to have a higher a* value (**Figure 11**). The major factor affecting

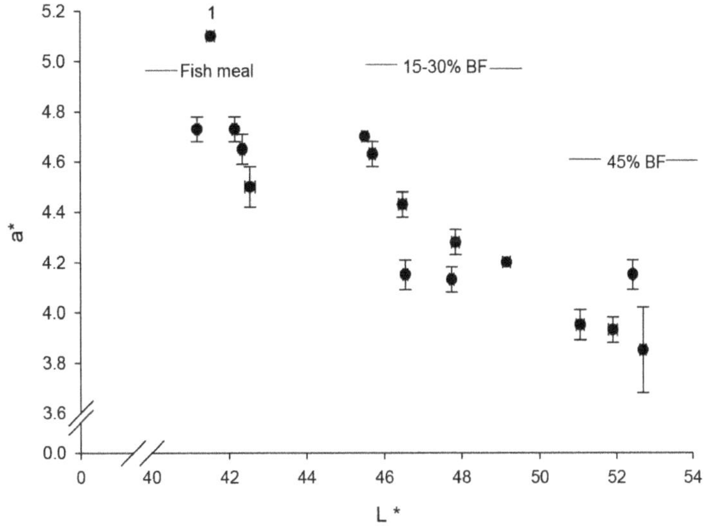

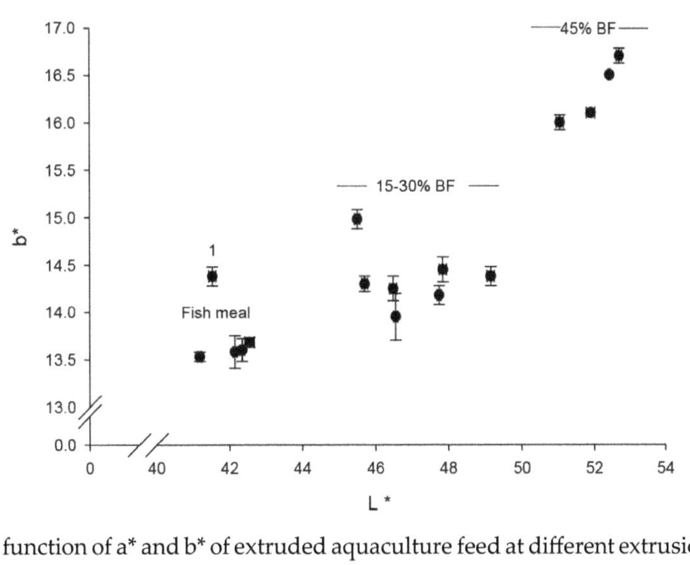

Figure 11. Lightness (L) in function of a* and b* of extruded aquaculture feed at different extrusion moisture with fishmeal and different concentrations of bean flour (*P. vulgaris*). BF = Bean flour; 1 = extruded at 150°C/22% moisture, and 15% BF; Vertical and horizontal bars indicate standard deviation of the mean (adapted from Rodríguez-Miranda et al., 2014).

L* and a* is the content of bean flour and fishmeal. Positive b* values (0–60) represent a yellow color and negative values (0—60) are blue. The samples with fishmeal and no bean flour (BF) have values ranging between 13.53 and 13.68 lower ($p < 0.05$) than the b* values for the extruded samples with bean flour, except the sample with 15% BF extruded at 150°C and 22% moisture. The extruded aquaculture feed with 15–30% of bean flour present a less gray color tending to yellowish with values between 14.18 and 14.98. For higher amounts of bean flour, the feed tends to be lighter in color and more yellowish (16.00–16.7). The differences in color are explained in part by the original color of the raw material; fishmeal tends to be more into the gray, less light color, while starch present in the bean flour also affects luminosity and color. The thermal process during extrusion also defines the final color of the feed. Different chemical reaction, such as milliard and caramelization interact to give the final luminosity and color of the product.

8. Effect of extrusion on bioactive compounds

The effect of extrusion moisture and temperature on the antioxidative capacity and bioactive compounds in bean/corn extrudates is shown in **Table 4**. Neither extrusion temperature nor extrusion moisture had an effect ($p > 0.05$) on the antioxidant activity [23]. The β-carotene, flavonoids, or polyphenols content was not affected ($p > 0.05$) by the extrusion temperature and moisture. An experimental central rotary design of second order was used for the extrusion experiment. The experiments were conducted in a single screw extruder with a temperature range from 142 to 198°C and 16.3 to 18.7% moisture content. Extrusion even at 192°C does not seem to affect the antioxidative activity, nor the concentration of the active compounds. Extrusion uses little processing time, which makes it an adequate way for food processing since it is not shedding for bioactive compounds.

	Intercept	Lineal		Quadratic		Interaction	R^2
	b_0	X_1	X_2	X12	X22	X_1X_2	
CUPRAC (μM Trolox equivalent/g)	−368.566	1.089	37.097	−2.83E–3	−1.060	−0.011	0.116
β-carotene (%)	−382.798	2.403	22.795	−3.79E–3	−0.324	−0.066	0.112
Flavonoids (mg Catechin equivalent/g)	−288.451	1.754	17.586	2.91E–3	−0266	−0.049	0.157
Polyphenols (mg Gallic acid equivalent/g)	−184.111	1.133	11.783	−2.07E–3	−0.209	−0.027	0.114

*Significant difference ($p < 0.05$). X_1 = temperature [°C], X_2 = moisture content [%].

Table 4. Regression coefficients of the response surface model of extruded bean 60%-bean/40%-corn extrudates (Delgado-Licon et al. [23]).

9. Extruded cottonseed meal

The use of cottonseed meal (CSM) in extruded snacks can double the amount of protein with just an increase of 10% CSM [9]. The protein concentration of an extruded snack can enhance from 6.4 to 12.8% when 10% CSM was added to the formulation. **Table 1** shows the protein content of different agricultural by-products and their chemical composition.

The difference in chemical composition changes the chemical and physical structure, the extrusion properties, and the functional properties of the final product. **Figure 12** shows two extruded samples. **Figure 12a** illustrates the structure of extruded corn masa, with a low protein content and a high starch content. It can be seen how layers of starch are built to provide expansion to the extruded product. On the other hand, **Figure 12b** shows the matrix of extruded cottonseed meal, which has 12.8% of protein. The flat, homogenous layers are gone, and a more irregular structure is present. It appears as if the protein breaks up the continuous starch structure and builds a less homogenous texture. The lower concentration of starch and the presence of more protein produce a more compact structure, which has a Lower Expansion Index and a harder crispier structure. The increase of protein content in extruded corn/cottonseed meal products reduces ($p < 0.05$) the physical and functional properties of the end-product; the Expansion Index, the water activity, and the water absorption and solubility indices decrease ($p < 0.05$) with protein cotton increase. As Expansion Index decreases the hardness of the extruded products increases. It is not only the low starch concentration that lowers the Expansion Index of the extruded products, but the matrix composition and structure also determines the hardness and Expansion Index of the final product. **Figure 12** shows how proteins produce a more compact irregular structure in cottonseed meal (CSM) extruded products. When CSM is extruded in a single screw extruder, an increase in CSM negatively affects ($p < 0.05$) the Expansion Index, because of the presence of protein. CSM also decreases ($p < 0.05$) the water activity, Water Absorption Index, and the Water Solubility Index of the extruded product [9]. In aquaculture, Low Water Solubility and Low Water Absorption Indices are most likely to be preferred, rather than high values. The extruded feed requires stability in an aqueous environment to assure that the fish or shrimp can have time to consume it. Stability of the extrudates also helps to reduce water turbidity and pollution. Low Water Solubility and Water Absorption Indices have a positive effect on the quality of aquaculture feed. On the other hand, a lower water activity (a_w) is related to a longer shelf life of the feed.

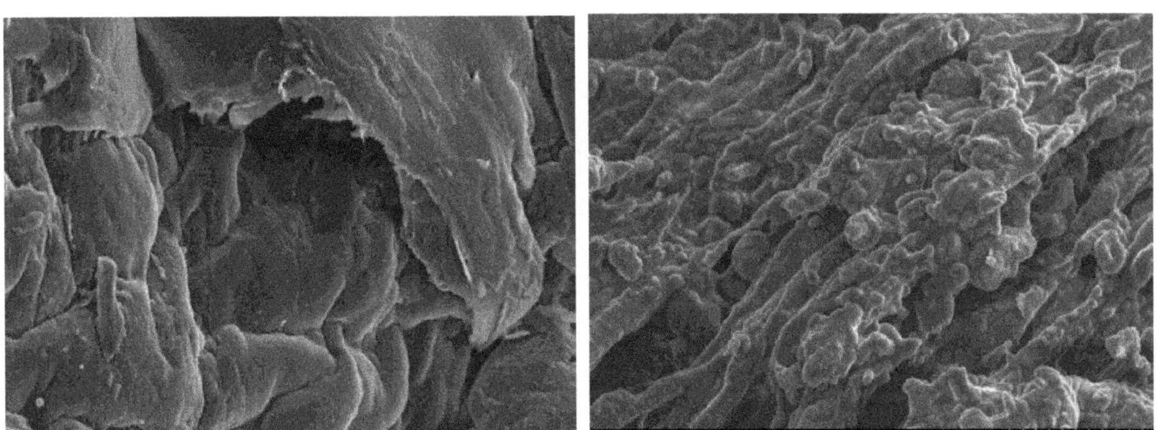

a b

Figure 12. Scan microscopy of extruded (a) corn masa and (b) cottonseed meal (reprinted with permission from Reyes-Jaquez et al. [9]).

Extrusion shows restructuring of cottonseed meal. **Figure 13a** shows a heterogeneous structure before extrusion and a homogenous structure after extrusion (**Figure 13b**). Lambda scan microscopy of extruded products shows different scans between samples with and without cottonseed meal (**Figure 14**). The extruded samples with cottonseed meal show a second pick at about 670 nm (**Figure 14b**), which is not shown in the samples without cottonseed meal (**Figure 14a**).

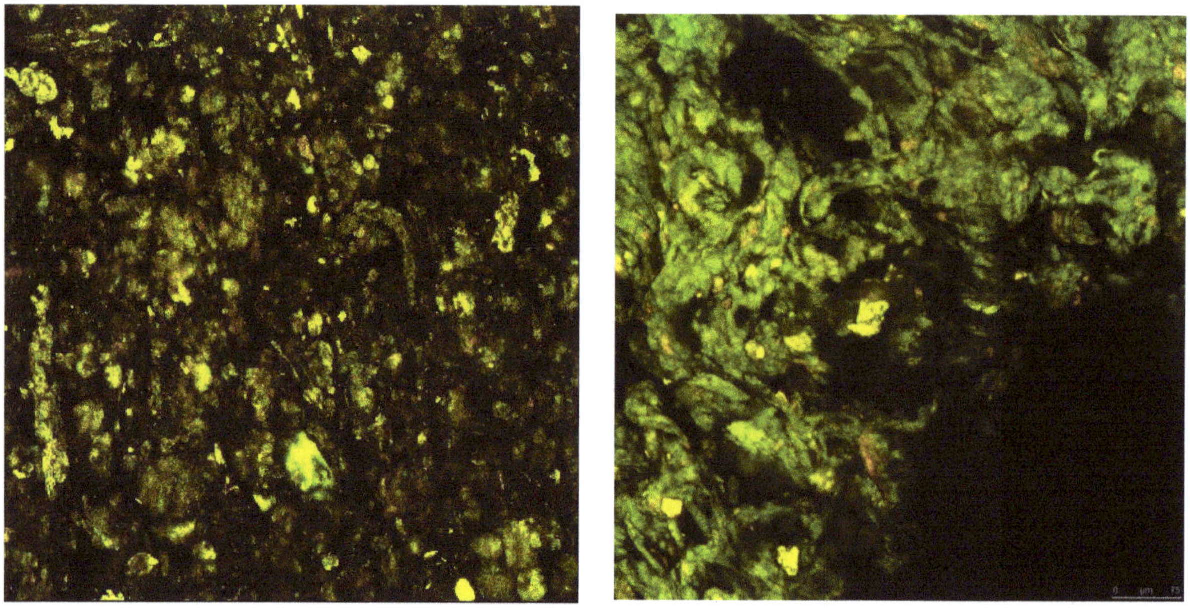

a b

Figure 13. Confocal microscopy of (a) not extruded and (b) extruded cottonseed meal (reprinted with permission from Reyes-Jaquez et al. [9]).

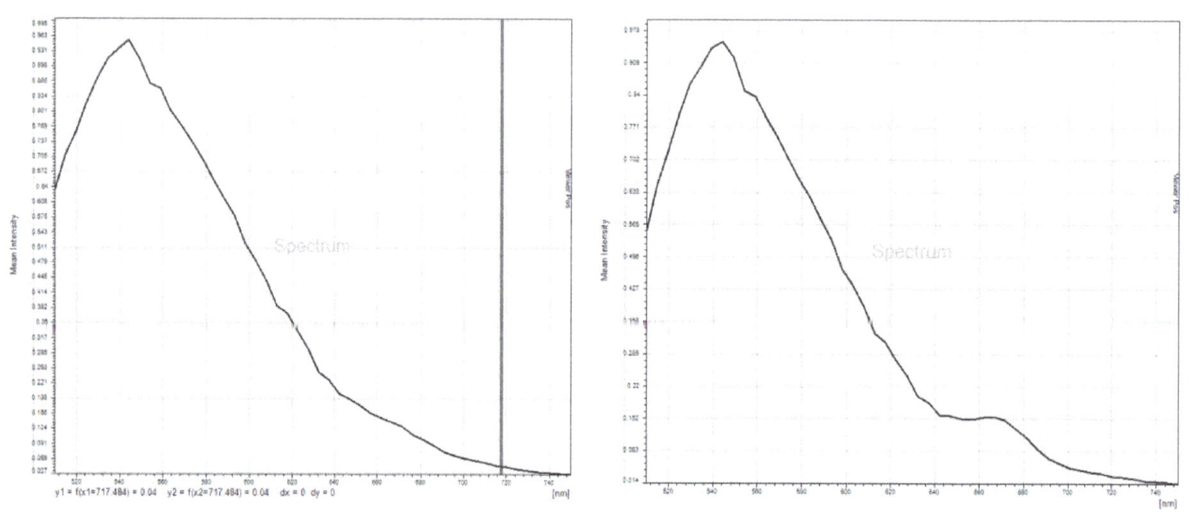

a b

Figure 14. Lambda scan microscopy of extruded (a) corn masa and (b) corn masa/cottonseed meal (reprinted with permission from Reyes-Jaquez et al. [9]).

Acknowledgements

This work was supported in part by the USDA National Institute of Food and Agriculture, Hatch project 1010849. This work was also supported in part by the Cotton Incorporated, USA project 126758.

Author details

Efren Delgado* and Damian Reyes-Jaquez

*Address all correspondence to: edelgad@nmsu.edu

Food Science and Technology, Department of Family and Consumer Sciences, New Mexico State University, New Mexico, USA

References

[1] FAO. The State of the World Fisheries and Aquaculture 2012. Rome: FAO; 2012

[2] USDA NASS. Census of aquaculture (2013). 2014;**3**:9

[3] FAO. FAO Global Aquaculture Production Database Updated to 2013 – Summary Information [Internet]. 2015. Available from: http://www.fao.org/fishery/statistics/en

[4] Tacon AGJ, Metian M. Global overview on the use of fish meal and fish Oil in industrially compounded aquafeeds: Trends and future prospects. Aquaculture. 2008;**285**:146-158

[5] Shen G, Fan X, Yang Z, Han L. A feasibility study of non-targeted adulterant screening based on NIRM spectral library of soybean meal to guarantee quality: The example of non-protein nitrogen. Food Chemistry. 2016;**210**:35-42

[6] Clive J. Global Status of Commercialized Biotech/GM Crops: 2012. Ithaca, NY: ISAAA; 2012. ISAAA Brief No. 44

[7] Atienzo-Lazos M, Delgado E, Ochoa-Martínez A, Aguilar-Palazuelos E, Martínez BF, Ramirez-Wong B, Gallegos-Infante A, Medrano-Roldan H, Solis-Soto A. Effect of moisture and temperature on the functional properties of composite flour extrudates from beans (Phaseolus vulgaris) and nixtamalized Corn (Zea mays). Journal of Animal Production Advances. 2011;**1**(1):9-20

[8] Delgado E, Vences-Montano MI, Rochas-Guzman N, Rodrigues-Vidal A, Herrera-Gonzalez SM, Medrano-Roldan H, Solis-Soto A. Inhibition of the growth of rats by extruded snacks from bean (Phaseolus vulgaris) and corn (Zea mays). Emirates Journal of Food and Agriculture. 2012;**24**(3):255-263

[9] Reyes Jáquez D, Casillas F, Flores N, Cooke P, Delgado Licon E, Solís Soto S, Andrade González I, Carrete Carreón FO, Medrano Roldán H. Effect of glandless cottonseed meal content on the microstructure of extruded corn-based snacks. Advances in Food Sciences. 2014;**36**(3):125-130

[10] Yu X, Yuan F, Fu X, Zhu D. Profiling and relationship of water-soluble sugar and protein compositions in soybean seeds. Food Chemistry. 2016;**196**:776-782

[11] Reyes-Jáquez D, Casillas F, Flores N, Andrade-González I, Solís-Soto A, Medrano-Roldán H, Carrete F, Delgado E. The effect of glandless cottonseed meal content and process parameters on the functional properties of snacks during extrusion cooking. Food and Nutrition Sciences. 2012;**3**:1716-1725

[12] Gerzhova A, Mondor M, Benali M, Aider M. Study of total dry matter and protein extraction from canola meal as affected by the pH, salt addition and use of zeta-potential/turbidimetry analysis to optimize the extraction conditions. Food Chemistry. 2016;**201**:243-252

[13] Gui D, Liu W, Shao X, Xu W. Effects of different dietary levels of cottonseed meal protein hydrolysate on growth, digestibility, body composition and serum biochemical indices in crucian carp (*Carassius auratus* gibelio). Animal Feed Science and Technology. 2010;**156**:112-120

[14] Mujahid A, Abdullah M, Barque AR, Gilani AH. Nutritional value of cottonseeds and its derived products: I. Physical fractions and proximate composition. Asian-Australasian Journal of Animal Science. 2000;**13**:348-355

[15] Khattab RY, Arntfield SD. Functional properties of raw and processed canola meal. LWT - Food Science and Technology. 2009;**42**:1119-1124

[16] Spiehs MJ, Whitney MH, Shurson GC. Nutrient database for distiller's dried grains with solubles produced from new ethanol plants in Minnesota and South Dakota. Journal of Animal Science. 2002;**80**:2639-2645

[17] Rodríguez-Miranda J, Gomez-Aldapa CA, Castro-Rosas J, Ramírez-Wong B, Vivar-Vera MA, Morales-Rosas I, Medrano-Roldan I, Delgado E. 2014. Effect of extrusion temperature, moisture content and screw speed on the functional properties of aquaculture balanced feed. Emirates Journal of Food and Agriculture. 2014;**26**(8):659-671

[18] Rodríguez-Miranda J, Ramírez-Wong B, Vivar-Vera MA, Solís-Soto A, Gómez-Aldapa CA, Medrano-Roldan H. Delgado E. Effect of bean flour concentration (*Phaseolus vulgaris* L.), moisture content and extrusion temperature on the functional properties of aquafeedS. Revista Mexicana de Ingeniería Química. 2014;**13**(3):649-663

[19] Rodríguez-Miranda J, Delgado-Licon E, Ramírez-Wong B, Solís-Soto A, Vivar-Vera MA, Gómez-Aldapa CA, Medrano-Roldán H. Effect of moisture, extrusion temperature and screw speed on residence time, specific mechanical energy and psychochemical properties of bean flour and soy protein aquaculture feeds. Journal of Animal Production Advances. 2012;**2**(1):65-73

[20] Rodríguez-Miranda J, Delgado-Licon E, Hernández-Santos B, Reyes-Jaquez D, Aguilar-Palazuelos E, Medrano-Roldan H, Navarro-Cortez RO, Castro-Rosas J, Gómez-Aldapa CA. The effect of pregelitinized potato starch on the functional properties of an extruded aquafeed. Journal of Animal Production Advances. 2012;**2**(7):335-344

[21] Rodríguez-Miranda J, Reyes-Jáquez D, Delgado E, Ramirez-Wong B, Esparza-Rivera JR, Solís-Soto A, Vivar-Vera MA, Medrano-Roldán H. Partial substitution of bean (Phaseolus vulgaris) flour for fishmeal in extruded diets for rainbow trout (Oncorhynchus mykiss): Effects on yield parameters. Iranian Journal Fisheries Sciences. 2016;**15**(1):206-220

[22] Morales AE, Cardenas G, de la Higuera M, Sanz A. Effects of dietary protein source on growth, fed conversión and energy utilization in rainbow trout, Onchorhynchus mykiss. Aquaculture. 1994;**124**:117-126

[23] Delgado-Licon E, Martinez Ayala AL, Rocha-Guzman N, Gallegos-Infante JA, Atienzo-Lazos M, Drzewiecki J, Martínez-Sánchez CE, Gorinstein S. Influence of extrusion on the bioactive compounds and the antioxidant capacity of the bean/corn mixtures. International Journal of Food Sciences and Nutrition. 2008;**1**:1-11

10

Extrusion Processing of Ultra-High Molecular Weight Polyethylene

Haichen Zhang and Yong Liang

Abstract

Ultra-high molecular weight polyethylene (UHMWPE) is a unique thermoplastic polymer with excellent performances. It has ultra-high molecular weight and extreme rheological behaviour, which make it a worldwide challenge to process UHMWPE continuously with little or without processing aids. Although the polymer processing technology has been increasingly maturated, it still cannot carry out the industrialized production efficiency by conventional processing methods and apparatus at present. In this chapter, we review the progress of extrusion processing technology for UHMWPE, including ram extrusion, single screw extrusion, twin screw extrusion and novel extrusion technology based on extensional rheology. By summarizing of these processing technologies, a basic framework of the processing principles and methods for UHMWPE is clearly presented. It is helpful for us to understand the processing characteristics and methods for such thermoplastic polymer with ultra-high molecular weight.

Keywords: ultra-high molecular weight polyethylene, extrusion processing, melting mechanism

1. Introduction

Ultra-high molecular weight polyethylene (UHMWPE) is a unique thermoplastic polymer possessing outstanding physical and mechanical performances such as good wear and corrosion resistance, low coefficient of friction, high impact strength at cryogenic temperatures, good environmental stress-cracking resistance, non-toxic and acceptable biocompatibility [1, 2]. Because of these excellent properties, UHMWPE has been widely used in many applications including mining, transportation, military industries, biomedical engineering, sports and livelihood projects [3–6].

UHMWPE is a special kind polyethylene. Except it, the polyethylene family also includes linear low density polyethylene (LLDPE), low density polyethylene (LDPE), high density polyethylene (HDPE) and cross-linked polyethylene (XLPE) [7–9]. These polyethylenes are synthesized with ethylene as monomers. By using different catalysts and synthetic process conditions, a variety of polyethylene with different structures, density and properties can be synthesized, such as LLDPE, LDPE and HDPE. In addition, cross-linked polyethylene can be obtained by the crosslinking reaction during or after the synthesis reaction process. UHMWPE is an unbranched linear polyethylene and can be synthesized with Ziegler-Natta or metallocene catalyst under the low-pressure polymerization process conditions [10, 11]. UHMWPE has extremely high molecular weight of up to several millions g/mol, which is several times larger than HDPE with a molecular weight of 200,000 g/mol. Actually, the precise molecular weight of UHMWPE is too high to be measured directly by conventional means, and it must be inferred by its intrinsic viscosity as an alternative approach.

In general, molecular weight of polymer has significant effects on its condensed structure, chemical performance, mechanical properties and processability. It is obvious that the excellent characteristics of UHMWPE are largely benefited from its extremely high molecular weight. The acceptable reasons are related to the enhanced intermolecular interactions and intensive chain entanglements. With increasing molecular weight, van der Waals force between macromolecules would be strengthened, and the ultra-long molecule chains of UHMWPE are prone to become entangled and to form intensive physical entanglements [12, 13]. These factors synthetically result in the significant improvement of mechanical strength, abrasion resistance, chemical stability, and so on.

Even if the ultra-high molecular weight has brought superior performances for UHMWPE, the extremely high molecular weight also brings great challenges for its processing and forming such as poor dissolution, extremely high melt viscosity and poor melt flowability [14–16]. The long chain movement of UHMWPE can be restricted by the strong constraint effect due to the enhanced intermolecular interactions and dense physical entanglements, resulting in improving the solvent resistance and solution or melt viscosity. This result is not a good thing for processing and forming of UHMWPE, including solution method and melting method.

For most thermoplastic polymers, there are a variety of processing and moulding methods, such as extrusion, injection, compression and casting. By these processing technologies, raw materials can be fabricated into polymer products with given profiles and appropriate properties such as large products with sufficient strength or micro/nano parts with high precision. However, it is very difficult to process nascent UHMWPE via conventional batch processing methods. UHMWPE extrusion products (pipes, sheets and bars) only can be extruded by large trust extruder with special screw structures, simultaneously adding a large amount of organic compounds as lubricants. According to the literature, even with improved upgrade screw extruder, it is almost impossible to extrude UHMWPE without any processing aids. The most straightforward reason is the extremely high melt viscosity and low melt flowability [16–19]. It is also very difficult to carry out injection moulding due to the poor fluidity of UHMWPE. Even though Huang et al. [20, 21] have obtained injection parts of UHMWPE blend containing 90 wt% commercial UHMWPE and 10 wt% ultra-low molecular weight polyethylene(ULMWPE), with a modified injection moulding technique named as oscillation shear injection moulding, pure UHMWPE parts with profiled surfaces still cannot manufactured directly by injection moulding. Thus, multistep processing method is usually used to prepare UHMWPE parts with complex profiles. For instance, in order to manufacture artificial knee joints, nascent UHMWPE powder particles must first be moulded into primary products with square or cylindrical profiles by compression moulding or plunger extrusion, and then artificial knee joints were fabricated by turning from primary products [1].

As a matter of fact, the problem of UHMWPE processing is always a worldwide challenge for material engineers and researchers from past to present. The efficient and easier processing solutions for UHMWPE could not be put forward over the past few decades. It is still a main constraint for promoting the development and application of UHMWPE at present. Actually, UHMWPE extrusion processing technique is the most likely to first achieve industrial production. The objective of this chapter is to review the progress of extrusion processing technique for UHMWPE, as a basis for understanding the processing principles and methods for such thermoplastic polymer with ultra-high molecular weight.

2. Flow characteristics of UHMWPE

2.1. Structure and thermal properties

As previously described, UHMWPE is synthesized with ethylene as monomers and has same repeat units as other common polyethylene. The single molecular chains of UHMWPE can consist of as many as hundreds of thousands repeat units. Because of the internal energy, the molecular chain could become mobile at elevated temperatures. Considering the ultra-long length and movement, a single molecular chain of UHMWPE is like a moving string over a kilometre long and tends to be tangled gradually.

As the temperature decreases, the molecular chain has a trend to rotate around the C-C bonds and create chain folds, reaching a new energy equilibrium state at a lower temperature. When cooled below the melt temperature of polymer, the activity of molecular chain is reduced, and the folded chains begin to form crystalline lamellae, which is the local ordered sheet-like regions. These lamellae gradually grow and accumulate in order to form crystalline regions of polymers. It is well known that there is almost no complete crystallization for polymer. In other words, there are always disordered regions in polymer, namely amorphous regions. Although the molecular chain of polyethylene is regular and flexible, not all segments of polyethylene chain can arrange into the ordered regions to form lamellae and crystalline regions. The lamellae are embedded within amorphous regions and may communicate with surrounding lamellae by tie molecules. Kurtz [1] has clearly observed the crystalline morphology of UHMWPE by using transmission electron microscopy (TEM), showing the composite nature of UHMWPE as an interconnected network of amorphous and crystalline regions. It can be clearly seen that UHMWPE is a typical semi-crystalline polymer as ordinary polyethylene.

Despite the close crystallinity, the morphology of nascent UHMWPE powder particles is quite different from that of HDPE. Obviously, from scanning electron microscope (SEM) images as shown in **Figure 1**, the individual nascent powder particle of UHMWPE is composed of several secondary particles. There are many orientation fibres as connections between secondary particles. Compared with other dense parts in the individual particle, the orientation fibres are loose. These fibres as weak sections in powder particles may be first melted when the temperature rises to the melting point.

The values of crystallinity degree and melting point of nascent UHMWPE with different molecular weight are listed as **Table 1**. Apparently, the crystallinity degree and melting point have a rising trend with increasing molecular weight. However, the values of crystallinity degree and melting point from the second scanning curves are all lower than that from the first scanning. The possible reason is that the crystallinity degree of polymer depends upon various factors, including molecular weight, processing conditions and environmental conditions (such as loading, flow field, ultrasonic, and so on) [22–25]. It is helpful to understand the melting mechanisms and melt flow process by realizing the melting process and crystallization process.

Thermal stability of polymer is another important factor, which must be considered in processing. In general, the decomposition temperature of HDPE or LDPE is below 400°C. But the decomposition temperature of UHMWPE usually exceeds 430°C probably because of the strong intermolecular interactions from the extremely high molecular weight. As the heating scanning curve of UHMWPE (GUR4120, Ticona) exhibited in **Figure 2**, the initial decomposition temperature is up to 450°C and the fastest degradation process occurred at about 480°C. Although it has higher decomposition temperature than other common polyethylene, the processing temperature for UHMWPE cannot be such high. Due to poor thermal conductivity, the accumulation of heat causes too high temperature in partial of polymer melt, and the thermal degradation in the melt state occurs in the processing temperature up to 300°C.

2.2. Flow characteristics

It is important to note that the melt flow characteristics of polymer usually determine the processing method and conditions. It also greatly affects the forming process and product quality. For instance, the phenomenon of melt breakup in extrusion is usually related to the critical shear rate, which is an important aspect of flow characteristics of polymer melt.

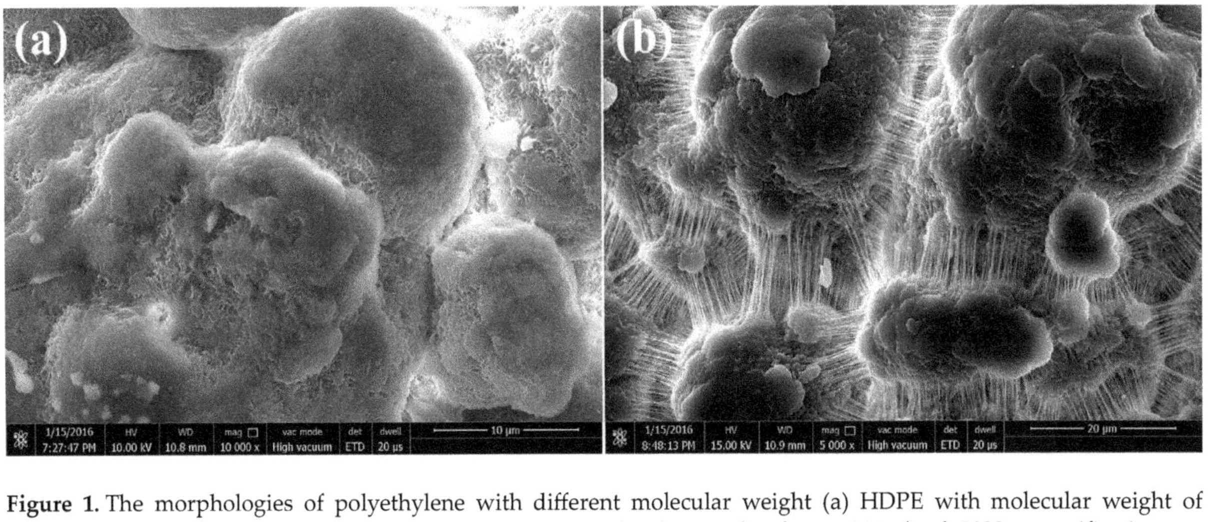

Figure 1. The morphologies of polyethylene with different molecular weight (a) HDPE with molecular weight of 600,000 g/mol, 10,000× magnification; (b) UHMWPE with molecular weight of 2,000,000 g/mol, 5000× magnification.

Samples	M_w (1×10^4 g/mol)	T_{m1} (°C)	T_{m2} (°C)	X_{c1} (%)	X_{c2} (%)
GHR8110	~60	130.2	128.5	71.1	68.3
145 M	~145	140.7	135.8	65.8	54.8
M1	150–200	143.6	136.4	69.3	54.0
M2	250–350	144.2	137.1	83.4	65.2
M3	~350	144.5	137.0	81.4	61.9
GUR4120	~500	144.5	136.4	78.5	58.9

Table 1. Melting point and crystallinity of various UHMWPE with different molecular weight from DSC curves (1—the first scanning, 2—the second scanning).

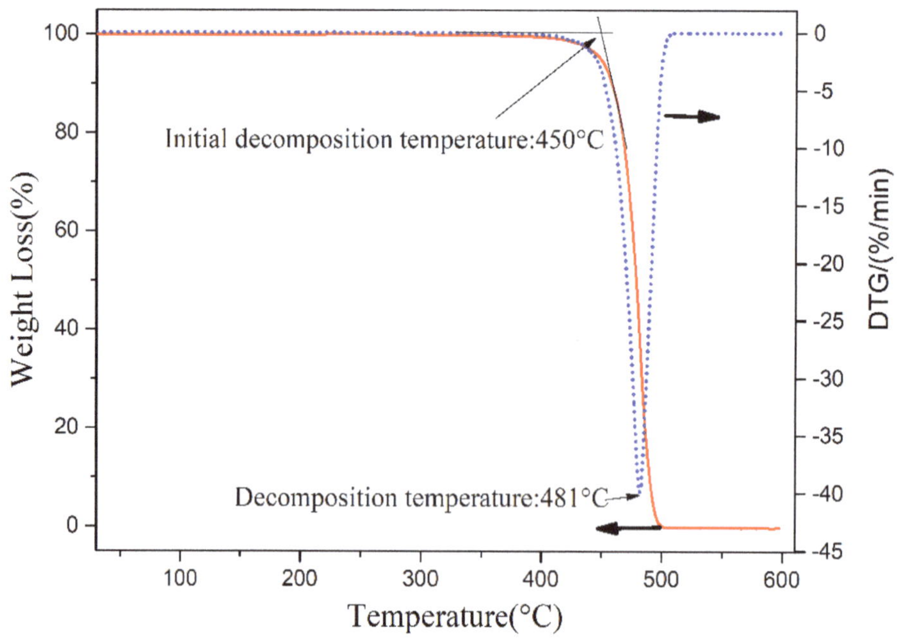

Figure 2. Thermogravimetric analysis (TGA) for UHMWPE.

For UHMWPE, its melting point is about 155°C, and the polymer powder particles can completely fuse into melt as temperature is higher than 200°C. However, the UHMWPE melt is rubberlike and has poor flowability even at the temperature much higher than melting point. Even if the temperature and load, respectively, were up to 250°C and 21.6 kg separately, the melt flow rate of UHMWPE was almost zero. Melt viscosity is an important parameter of melt flow. Actually, it is very difficult to determine directly the precise value for the melt viscosity of UHMWPE. But the trend in change of viscosity can be inferred with molecular weight by the following equations [16].

$$\eta = kM_w (M_w < M_{wc}) \quad (1)$$

$$\eta = kM_w^{3.4} (Mw > M_{wc}) \quad (2)$$

where k is a constant and η, M_w, and M_{wc} represent viscosity, average molecular weight, and critical relative average molecular weight, respectively.

Generally speaking, the molecular weight of UHMWPE should be higher than 1.5 million g/mol. Thus, the molecular weight of UHMWPE is much higher than the critical relative average molecular weight. Therefore, the viscosity can be evaluated by Eq. 2. For instance, the viscosity of UHMWPE with molecular weight of 3 million g/mol is about 2500 times higher than HDPE with molecular weight of 300,000 g/mol. According to the references, the melt viscosity of UHMWPE could be up to 10^8 Pa·s [26].

The extremely high viscosity would certainly result in the poor melt flowability and processability. For example, even if the nascent UHMWPE powder particles were processed with an internal mixer at 200°C for 20 min, the masterbatch still could not be dispersed evenly in UHMWPE melt. As displayed in **Figure 3a**, during the whole mixing process, the masterbatch only dispersed in a very narrow range along the rotation direction; however, it almost cannot disperse into the melt far away from the place where masterbatch is placed. Probably because the strong shearing effect along rotation direction makes materials exchanging significantly, and there is no obvious material exchange in the direction perpendicular to the rotation. This clearly demonstrates the limited molecular chain mobility and poor melt flowability of UHMWPE melt.

With the increase of temperature, UHMWPE melt will not enter the viscous flow state but to maintain a transparent rubberlike state. In fact, UHMWPE melt has no viscous flow state like HDPE or LDPE because its theoretical viscous flow temperature is higher than the decomposition temperature. In addition, it is easy to find out from the picture that the UHMWPE melt has ruptured and could not form uniform and continuous melt. The fundamental reason is that UHMWPE melt has a low critical shear rate of 10^{-3} s^{-1}. Therefore, the UHMWPE melt was easy to break up by strong shear effect with internal mixer. Although there exists a metastability processing window in the temperature range of 154°C and 157°C [27–29], the nascent UHMWPE cannot be processed by conventional or improved upgrade screw equipment without processing aids. Considered the complicated interfering factors of extremely high melt viscosity, low melt flow rate, low critical shear rate no viscous flow and so on, it is a huge challenge to process UHMWPE continuously and efficiently via screw equipment dominated by shear flow and methods based on viscous flow theory.

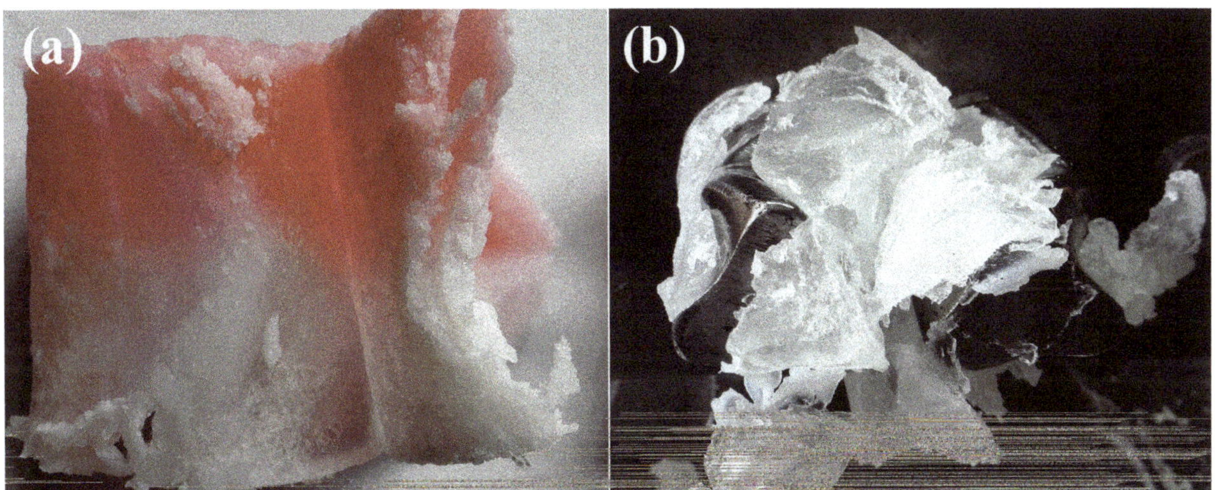

Figure 3. Melting state of UHMWPE processed with an internal mixer.

3. Extrusion processing and forming of UHMWPE

3.1. Ram extrusion of UHMWPE

The most common methods used to process UHMWPE powder particles into bulk products are compression moulding and ram extrusion [30, 31]. Compression moulding originated in Germany in the 1950s is a discontinuous process based on sintering without limitation of molecular weight and melt viscosity. Ram extrusion was developed by converters in the United States during the 1970s. It was a processing process, which can be considered as a continuous compression and sintering process. Taking into account the intermittent feeding and compaction, ram extrusion in the strict sense was a quasi-continuous process. Nevertheless, it still could consecutively produce sheets, pipes, bars and profiles, even the molecular weight of UHMWPE being up to 10 million g/mol. In contrast with single screw or twin screw extrusion, the ram extrusion only was affected slightly by molecular weight and melt viscosity.

Ram extruder consists essentially of a hopper, a feeding chamber with heating device, a horizontal reciprocating ram, a heated forming die, cooling and shaping apparatus. So it is easy to understand the ram extrusion process of feeding, compaction, melting and plasticizing, extruding, cooling, and moulding. UHMWPE powder particles are fed continuously into ram extruder and were heated at elevated temperatures. Within the extruder, the powder particles are consolidated and maintained under pressure by the ram, as well as by the back pressure from UHMWPE melt, which caused by frictional forces of the molten resin against the die wall surface. UHMWPE melt is extruded from heated forming die with specified shape. The thrust force to overcome the huge resistance originated from positive displacement movement of the reciprocating ram.

Although the ram extruder can manufacture UHMWPE products with good surface quality, the intermittent stamping process caused residual stresses inside the bulk. It is necessary for extrusion products to be annealed at elevated temperatures in order to remove residual stresses. The annealing process can also increase the crystallinity of the components, which is helpful to maintain the excellent mechanical performances of UHMWPE.

There are also some disadvantages for ram extruder to process UHWMPE, such as fluctuations in product quality, longer melt plasticization cycle, slow extrusion rate, and high energy consumption. Extruder with multiple plungers is used to reduce the pulsation frequency [32]. Several rams alternately compact the materials and push them forward. This process shortens the operation time between two compaction actions, making the whole process closer to a continuous process, which is conducive to reducing the fluctuations of processing process and product quality. Certainly, this requires more advanced control technology for the extrusion process.

3.2. Screw extrusion of UHMWPE

In fact, regardless of high energy consumption and low processing efficiency, compression and ram extrusion are the most suitable processing methods for thermoplastic polymer with extreme high molecular weight and melt viscosity. However, the cost and diversity of products must be considered in industrial production. Screw extrusion is quite popular with people in all conventional processing methods for UHWMPE because of continuous production process and well-compounded effect. The commonly used screw equipment for UHMWPE are single screw extruder and twin screw extruder.

3.2.1. *Single screw extrusion*

In 1939, Troester Machinery Company in Germany launched an extruder with length/diameter ratio of 10, marking the development and rise of modern single screw extruder. There are many types of single screw extruders, which are widely used in extrusion processing and moulding of polymers [33–35].

Despite various kinds of single screw extruders, they have similar functions for common thermoplastic polymers. After entering the barrel from the hopper, the material was gradually pushed to the head direction with rotation of the screw. Successively, the material passed through several functional areas of the extruder, including solid conveying section, melting section, and melt conveying section. Loose materials were compacted in solid conveying section and melted before reaching the melt conveying section, then the homogenized melt was squeezed out from heating die. For most thermoplastic polymers, they eventually became viscous fluid in the extrusion process. Various processing rheology theories and extrusion equipment are based on this fact.

However, many cases are very different for UHMWPE. Since the low friction coefficient of UHMWPE and metal, UHMWPE powder particles in the feeding section are easy to slide with the rotating screw, resulting in the difficulty for powder particles being pushed forward. On the other hand, the melt is like rubber without viscous flow, which means a poor flowability for UHMWPE melt. The extremely high resistance is easily established in compaction section and results in huge backpressure due to the extremely high melt viscosity and poor flowability. Then, the melted resins are easy to wrap in the screw and rotate with the screw, preventing UHMWPE melt to move forward. It can even cause the screw to break if the device is forced to run.

Obviously, conventional single screw extruder is almost powerless to process UHMWPE. Then, many new dedicated equipment with special internal structures were developed in order to overcome the difficulty of feeding and huge extrusion resistance [36]. For example, special single screw extruder with gradient grooves in the barrel was developed in 1971 to avoid the slippage phenomenon of UHMWPE in extrusion processing by Mitsui Petrochemical Company. Depth and width of grooves are gradually decreasing along the extrusion direction, which is favourable for establishing pressure. The principle of establishing pressure is similar to the advanced extrusion system of Institute of Plastics Processing (IKV) at RWTH Aachen University. The pressure peak of extrusion system usually appears at the end of the solid conveying section. Such extrusion system is good for improvement of the extrusion output and stability. Many other special extruders are developed to solve delivery problems for UHMWPE. However, such extruders only increase the coefficient friction between polymer and barrel and do not change the conveying mechanism, which resulting in the increases of wear of screw and barrel, drive load, and friction heat. Although the energy consumption is increasing, single screw extruder with large thrust screw and special screw structures for enhancing conveying capability is the most practical processing device for UHMWPE at present.

As we know, in extruding process, common thermoplastic polymers successively experience solid state, viscous flow state, and high elastic state from feeding section to heating die. However, UHMWPE only experiences two physical states inside the barrel, namely solid state and high elastic state. The rotating screw consecutively grabbed UHMWPE powder particles from the feed inlet and compact them into block. The block is like a solid plug and conveys spirally forward along the rectangular tunnel by the trust of screw flight. Even the block is completely melted in metering section, it still moves forward as a whole. In other words, the flow mode of UHMWPE is plug flow, and there is almost no material exchange during the extrusion process. Therefore, the melting process of UHMWPE is similar to that of it under the static action by compression loads.

Polymer melt can be extruded out the extrusion die and forms continuous extrudates with specified cross-sectional shape as the die. During this process, polymer melt passes through the convergent channel inside extruder head and produces a shear deformation. Within the appropriate extrusion processing window, polymer melt with good flowability will maintain as a continuous melt block after flowing through the convergent channel.

As the schematic diagram shown in **Figure 4**, UHMWPE melt has very low critical shear rate, and the extrusion process will produce significant fluctuations when the shear rate is more than 10^{-2} s^{-1}. Many facts have shown that pure UHMWPE melt is easy to fracture even the extrusion speed was less than 10 r/min. The possible reason is that UHMWPE melt is fractured by shearing effect from deformation, and the fractured melt could not quickly merge to be continuous melt block again due to the poor molecular mobility of UHMWPE.

Many measures have been taken to prevent the unstable extrusion process and to improve the surface quality of extrudates, for instance, extending the length of the parallel flow path behind the convergent channel, reducing the angle of convergence, and reducing the friction resistance by using lubricants. Nevertheless, UHMWPE productivity and production efficiency have not improved significantly.

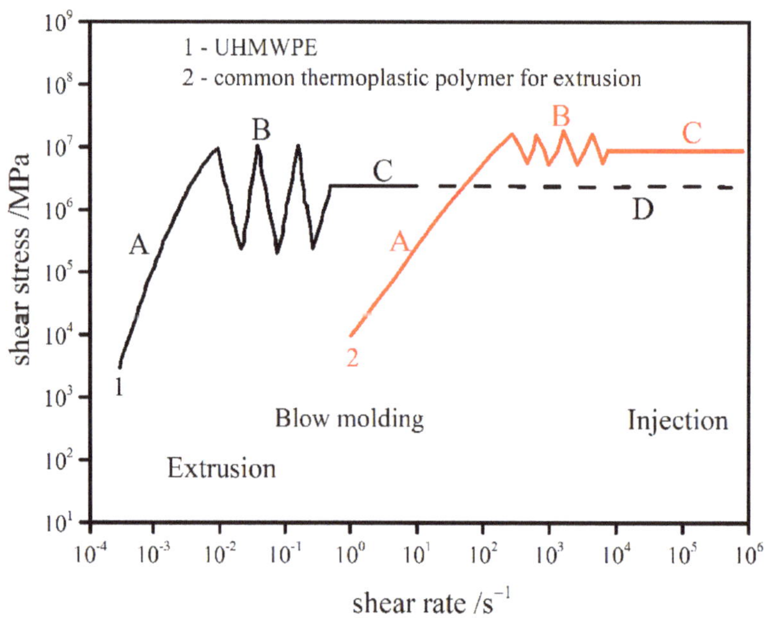

Figure 4. Schematic diagram of UHMWPE melt flow state during extrusion processing.

3.2.2. *Twin screw extrusion*

According to the relative rotation of two screws, twin screw extruders can be divided into counter rotating and co-rotating. The extruder of counter rotating has discontinuous channel that the spiral forward channel of screw is blocked by the screw flight of another screw with opposite rotating direction. For solid plug of UHMWPE, the continuous melt will be crushed into pieces by the strong shearing and mixing effect, and these crushed melt cannot quickly re-fuse to be a continuous melt block, causing significant instability extrusion.

Compared with counter rotating twin screw extruder, co-rotating twin screw extruder has a continuous channel to connect the feeding section and heating die, possessing good self-cleaning, and forcibly conveying capacity. The solid plug of UHMWPE can be conveyed forward along the continuous channel under positive displacement force. Such kind of extruder can effectively prevent material slippage and blockage.

At present, two types of twin-screw extruder are usually used to extrude UHMWPE in industrial production [20, 37, 38]. However, the difficult realities of low extrusion output, high energy consumption, and large driving load are still troubling people and hindering the development of UHMWPE processing.

3.2.3. *Novel extrusion process of UHMWPE*

As previously summarized, the nascent UHMWPE could be directly processed via compression moulding and ram extrusion. However, in most cases, processing aids should be required for screw extrusion even using an appropriative screw with special structures, and the excellent performances of UHMWPE would be damaged evidently. On the other hand, strong shearing action of screw extruder could make ultra-long molecular chains broken and even cause thermal degradation of UHMWPE.

Some new processing technologies are adopted to process UHMWPE in order to improve the extrusion output and reduce damage of material properties such as ultrasonic-assisted extrusion, gas-assisted extrusion, and near melting point extrusion process [39, 40]. Despite some progress, these extrusion processes still cannot extrude nascent UHMWPE without processing aids.

With regard to novel equipment for polymer processing, professor Qu and his research team [41–43] have independently developed novel extruders without screws such as vane extruder as shown in **Figure 5** [42] and eccentric rotor extruder as shown in **Figure 6** [43].

Compared with screw extruders, the rotor instead of screw was used in novel extruders and the rotor rotated eccentrically. The stator and rotor with special structures constitute continuous spatial path. The space volume between the rotor and stator periodically changes along the stator axial direction, which will make the volume of materials inside the spatial path change periodically along the axial direction, so that the materials were mainly subjected to stretch deformation. Put another way, the main flow field in such novel extruder was dominated by elongational flow field, and the materials were conveyed forward by positive displacement action.

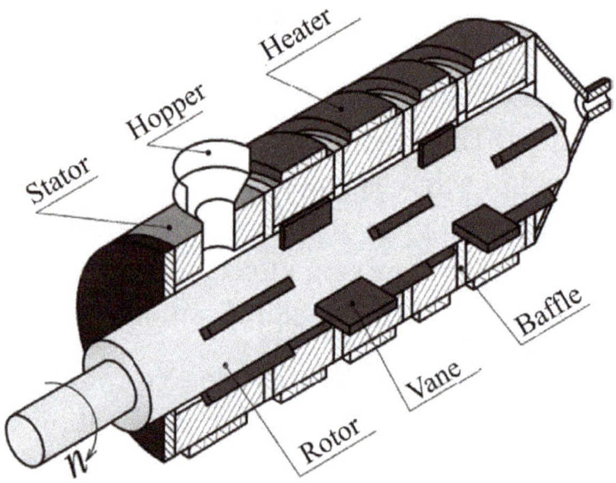

Figure 5. The schematic diagram of vane plasticizing and conveying system.

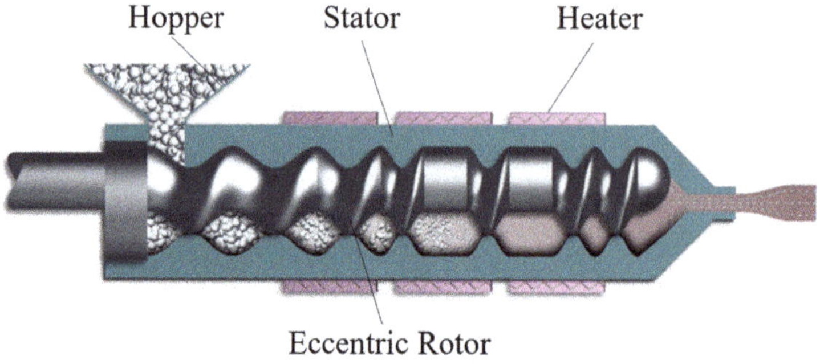

Figure 6. The schematic diagram of the eccentric rotor plasticizing and conveying system.

The eccentric rotor extruder has the same basic processing principle as vane extruder. Furthermore, eccentric rotor extruder has streamlined flow path, which is conducive to prevent thermal degradation. Nascent UHMWPE powder particles without any processing aids can be extruded directly by using an eccentric rotor extruder. The mechanical properties of the extrudates have almost been maintained as the unextruded samples. Since the material was conveyed by positive displacement force in the entire extrusion process, the energy consumption and extrusion output of UHMWPE were almost the same as that of HDPE. The melting of UHMWPE inside eccentric rotor extruder is a dynamic melting process because of the periodic variation of spatial path. However, the fluctuation of extrusion caused by the pulsating movement of eccentric rotor is one of the problems to be solved at present. Regardless of the current problems, the novel processing based on positive displacement delivery can significantly improve the production efficiency and quality of nascent UHMWPE.

4. Conclusions

The ultra-high molecular weight of UHMWPE has led to great difficulty in processing such as poor dissolution, extremely high melt viscosity, and poor melt flowability. The improved screw equipment with special structures can process UHMWPE with molecular weight of up to several hundred g/mol, but such modified upgrade equipment still cannot process the nascent UHMWPE without plasticizer and lubricants. The novel extrusion device of eccentric rotor extruder based on pulsating movement can directly process nascent UHMWPE without using any processing aids. Since the material is conveyed by positive displacement force in the entire extrusion process, the energy consumption and extrusion output of UHMWPE are almost the same as that of HDPE, showing small dependence on molecular weight. The melting process of UHMWPE inside eccentric rotor extruder is a dynamic process because of the periodic variation of spatial path.

Acknowledgements

The authors are very thankful to the Research Foundation for Talented Scholars Program of Foshan University (GG040946), Functional Polymer Materials Engineering Center Program of Foshan City (2016GA1062), and Engineering Technology Research Center Program of Educational Commission of Guangdong Province (2016GCZX008).

Author details

Haichen Zhang[1,3*] and Yong Liang[2]

*Address all correspondence to: haichen_zhang@163.com

1 School of Materials Science & Energy Engineering, Foshan University, Foshan, Guangdong, P.R. China

2 Changzhou Institute of Technology, Changzhou, Jiangsu, P.R. China

3 National Engineering Research Center of Novel Equipment for Polymer Processing, South China University of Technology, Guangzhou, Guangdong, P.R. China

References

[1] Kurtz SM. UHMWPE Biomaterials Handbook: Ultra High Molecular Weight Polyethylene in Total Joint Replacement and Medical Devices. Academic Press; 2009

[2] Muratoglu OK, Bragdon CR, O'Connor DO, Jasty M, Harris WH, Gul R, McGarry F. Unified wear model for highly crosslinked ultra-high molecular weight polyethylenes (UHMWPE). Biomaterials. 1999;**20**(16):1463-1470

[3] Deplancke T, Lame O, Cavaille J-Y, Fivel M, Riondet M, Franc J-P. Outstanding cavitation erosion resistance of ultra high molecular weight polyethylene (UHMWPE) coatings. Wear. 2015;**328**:301-308

[4] Veruva SY, Lanman TH, Isaza JE, MacDonald DW, Kurtz SM, Steinbeck MJ. UHMWPE wear debris and tissue reactions are reduced for contemporary designs of lumbar total disc replacements. Clinical Orthopaedics and Related Research. 2015;**473**(3):987-998

[5] Brach del Prever EM, Bistolfi A, Bracco P, Costa L. UHMWPE for arthroplasty: Past or future? Journal of Orthopaedics and Traumatology. 2009;**10**(1):1-8

[6] Wroblewski BM, Siney PD, Fleming PA. Penetration of UHMWPE Cup: Wear or Creep. In Charnley Low-Frictional Torque Arthroplasty of the Hip: Practice and Results. Cham: Springer International Publishing; 2016. pp. 263-265

[7] Kelly JM. Ultra-high molecular weight polyethylene*. Journal of Macromolecular Science, Part C: Polymer Reviews. 2002;**42**(3):355-371

[8] Bellare A, Spector M. The Polyethylene History in Total Knee Arthroplasty. Springer; 2005. pp. 45-50

[9] Bostrom MP, Bennett AP, Rimnac CM, Wright TM. The natural history of ultra high molecular weight polyethylene. Clinical Orthopaedics and Related Research. 1994; **309**:20-28

[10] Bae J-S, Kim T, Lee C. Synthesis of novel violet dyes for polyolefin fibers using N,N-dihexyl-2-methoxy-5-methylaniline and various diazo components. Fibers and Polymers. 2014;**15**(12):2466-2471

[11] Romano D, Ronca S, Rastogi S. A hemi-metallocene chromium catalyst with trimethylaluminum-free methylaluminoxane for the synthesis of disentangled ultra-high molecular weight polyethylene. Macromolecular Rapid Communications. 2015;**36**(3):327-331

[12] Chang B, Akil HM, Nasir RM, Nurdijati S. Mechanical and antibacterial properties of treated and untreated zinc oxide filled UHMWPE composites. Journal of Thermoplastic Composite Materials. 2011;**24**(5):653-667

[13] Marissen R. Design with ultra strong polyethylene fibers. Materials Sciences and Applications. 2011;**2**(05):319

[14] Rotzinger BP, Chanzy HD, Smith P. High strength/high modulus polyethylene: Synthesis and processing of ultra-high molecular weight virgin powders. Polymer. 1989;**30**(10): 1814-1819

[15] Fang L, Leng Y, Gao P. Processing and mechanical properties of HA/UHMWPE nanocomposites. Biomaterials. 2006;**27**(20):3701-3707

[16] Wood W. Processing, Wear, and Mechanical Properties of Polyethylene Composites Prepared with Pristine and Organosilane-Treated Carbon Nanofibers. Washington State University; 2012

[17] Ronca S, Igarashi T, Forte G, Rastogi S. Metallic-like thermal conductivity in a lightweight insulator: Solid-state processed ultra high molecular weight polyethylene tapes and films. Polymer. 2017;**123**:203-210

[18] Robert D, Hufen J, Lüdtke K, Rinker B, Ehlers J. Process for producing high molecular weight polyethylene. 2015. US Patents

[19] Spencer LP, Kirschner JM. Polymerization processes for high molecular weight polymers. 2017. US Patents.

[20] Huang Y, Xu J, Zhang Z, Xu L, Li L, Li J, Li Z. Melt processing and structural manipulation of highly linear disentangled ultrahigh molecular weight polyethylene. Chemical Engineering Journal. 2017;**315**:132-141

[21] Huang Y, Zhang Z, Xu J. Simultaneously improving wear resistance and mechanical performance of ultrahigh molecular weight polyethylene via cross-linking and structural manipulation. Polymer. 2016;**90**:222-231

[22] Guan C, Yang H, Li W, Zhou D, Xu J, Chen Z. Crystallization behavior of ultra-high molecular weight polyethylene/polyhedral oligomeric silsesquioxane nanocomposites prepared by ethylene in situ polymerization. Journal of Applied Polymer Science. 2014;**131**(19):40847 (9 pp.)–40847 (9 pp.)

[23] Deplancke T, Lame O, Rousset F, Aguili I, Seguela R, Vigier G. Diffusion versus cocrystallization of very long polymer chains at interfaces: Experimental study of sintering of UHMWPE nascent powder. Macromolecules. 2014;**47**(1):197-207

[24] Liu S, Zhao B, He D. Crystallization and microporous membrane properties of ultrahigh molecular weight polyethylene with dibenzylidene sorbitol. Journal of Applied Polymer Science. 2014;**131**(17) 40706 (8 pp.)–40706 (8 pp.)

[25] Doshi BN, Ghali B, Godleski-Beckos C, Lozynsky AJ, Oral E, Muratoglu OK. High pressure crystallization of vitamin E-containing radiation cross-linked UHMWPE. Macromolecular Materials and Engineering. 2015;**300**(4):458-465

[26] Hikosaka M, Tsukijima K, Rastogi S, Keller A. Equilibrium triple point pressure and pressure-temperature phase diagram of polyethylene. Polymer. 1992;**33**(12):2502-2507

[27] Ivan'kova E, Myasnikova L, Marikhin V, Baulin A, Volchek B. On the memory effect in UHMWPE nascent powders. Journal of Macromolecular Science, Part B. 2001; **40**(5):813-832

[28] Fang LM, Gao P, Cao XW. Temperature window effect and its application in extrusion of ultrahigh molecular weight polyethylene. Express Polymer Letters. 2011;**5**(8):674-684

[29] Gai J, Zuo Y. Metastable region of phase diagram: Optimum parameter range for processing ultrahigh molecular weight polyethylene blends. Journal of Molecular Modeling. 2012;**18**(6):2501-2512

[30] Blunn G, del Preva EB, Costa L, Fisher J, Freeman M. Ultra high molecular-weight polyethylene (UHMWPE) in total knee replacement: Fabrication, sterilisation and wear. Bone & Joint Journal. 2002;**84**(7):946-949

[31] Kurtz SM, Mazzucco D, Rimnac CM, Schroeder D. Anisotropy and oxidative resistance of highly crosslinked UHMWPE after deformation processing by solid-state ram extrusion. Biomaterials. 2006;**27**(1):24-34

[32] Saghafi HR, Naderifar A, Gerami S, Farasat A. Performance evaluation of viscosity characteristics of enhanced preformed particle gels (PPGS). Iranian journal of chemistry and chemical engineering. 2016;**35**(3)

[33] Shi D, Liu C, Qin J. Extrusion pressure analysis of powder state material of non-plug solid conveying in feeding section of single screw extruder. In: Proceedings of the 2015 International Conference on Materials, Environmental and Biological Engineering. 2015;**10**:334-338

[34] Wang F, Liu L, Xue P, Jia M. Crystal structure evolution of UHMWPE/HDPE blend fibers prepared by melt spinning. Polymer. 2017;**9**(3):96

[35] Schaeffer G, Hoffarth D. Feed bushing for single-screw extruders. 1984. US Patents.

[36] Davis BA, Gramann PJ, Noriega E, Del PM, Osswald TA. Grooved feed single screw extruders—Improving productivity and reducing viscous heating effects. Polymer Engineering & Science. 1998;**38**(7):1199-1204

[37] Tang W, Santare MH, Advani SG. Melt processing and mechanical property characterization of multi-walled carbon nanotube/high density polyethylene (MWNT/HDPE) composite films. Carbon. 2003;**41**(14):2779-2785

[38] Rocha LFM, Cordeiro SB, Ferreira LC, Ramos FJH, de Fátima Marques M. Effect of carbon fillers in ultrahigh molecular weight polyethylene matrix prepared by twin-screw extrusion. Materials Sciences and Applications. 2016;**7**(12):863

[39] Liu G, Li H. Extrusion of ultrahigh molecular weight polyethylene under ultrasonic vibration field. Journal of Applied Polymer Science. 2003;**89**(10):2628-2632

[40] Whitehouse C, Liu ML, Gao P. Cold extrusion and in situ formation of self-blends of UHMWPE: Part 1. Processability and thermal characterization. Polymer. 1999; **40**(6):1421-1431

[41] Zhang H, Huang J, Yang L, Chen R, Zou W, Lin X, Qu J. Preparation, characterization and properties of PLA/TiO_2 nanocomposites based on a novel vane extruder. RSC Advances. 2015;**5**(6):4639-4647

[42] Qu J, Zhang G, Chen H, Yin X, He H. Solid conveying in vane extruder for polymer processing: Effects on pressure establishment. Polymer Engineering & Science. 2012;**52**(10):2147-2156

[43] Wu T, Yuan D, Qiu F, Chen R, Zhang G, Qu J. Polypropylene/polystyrene/clay blends prepared by an innovative eccentric rotor extruder based on continuous elongational flow: Analysis of morphology, rheology property, and crystallization behavior. Polymer Testing. 2017;**63**(Supplement C):73-83

11

Application of Open-die Warm Extrusion Technique in Spur Gear Manufacturing

Wei Wang and Jun Zhao

Abstract

The open-die warm extrusion technique is recommended for spur gear manufacturing. This forming technique is systematically researched by using numerical simulation analysis and physical experiments. The lubricating condition, entrance angle, and initial blank size are determined as the crucial factors on the forming quality. The influence of each factor on this technology is fully understood and ascertained. The reasons for causing the forming defect in insufficient sections are analyzed and the die structure and extrusion speed are optimized by using the response surface method (RSM) for defects control and improving the forming quality. Furthermore, the improved process, "Variable Contour Two-Step Warm Extrusion," is presented in order to obtain good forming results in a poor lubricating condition.

Keywords: open die, warm extrusion, spur gear

1. Introduction

Traditional manufacturing method of spur gears is tooth machining from a round billet, which has the disadvantage of low material utilization. Also, the comprehensive mechanical performance is weakened as a result of metal fibers having been cut off. With the growing severity of energy crisis and environment problems, a new technology is urgently desired to be developed to replace the traditional manufacturing method. Spur gears manufacturing by forming technique can effectively overcome the above shortcomings. This is also beneficial for light weight spur gears because of retained metal fibers and increased power density [1]. Forging is one of the forming technologies first recommended for spur gears manufacturing. Closed-die forging [2, 3], floating die design [4–6], divided flow method [7, 8], and a series of improved technologies have been developed [9–12]. Throughout its development history, a key constraint of the spur gear forging is the contradictory relationship between the forming load and the forming precision.

Compared with the spur gear forging process, open-die extrusion has the advantage of lower forming load, which has the great significance to prolong die life [13, 14]. Pinions and splined shafts with small modules have been manufactured by extrusion technique [15, 16]. Open-die warm extrusion technique is a combination of the warm forming and the extrusion process, which has a series of obvious advantages: (1) compared with hot forming, this technology can effectively reduce the metal oxidation and decarburization and prevent overheating, overburning, and grain growth, (2) compared with cold forming, this technology can manufacture large module gears and avoid material cracking, and (3) compared with gear forging process, this technology needs smaller forming load so that it can overcome the premature die failure [17, 18].

Based on the advantages of the open-die warm extrusion, it has been developed by authors to manufacture spur gears [19, 20]. A comprehensive study of the open-die warm extrusion process of spur gears is carried out. The influence factors of this technology are fully explored. The RSM is adopted to control the forming defects and improve the forming quality. In addition, an improved process is presented.

2. Process model and forming principle

2.1. Process model

The process model of open-die warm extrusion of spur gear is illustrated in **Figure 1**. The punch is designed as a gear-like structure in order to avoid a big burr using a round punch or a big forming force using a whole tooth punch. The die includes an entrance section and a forming section. The entrance section has a correlative entrance angle (θ) and height (h) and the forming section has toothed contour. High-precision gears need grinding or honing, so the forming section has a contour offset Δ (0.30 mm) as machining allowance. In the extrusion

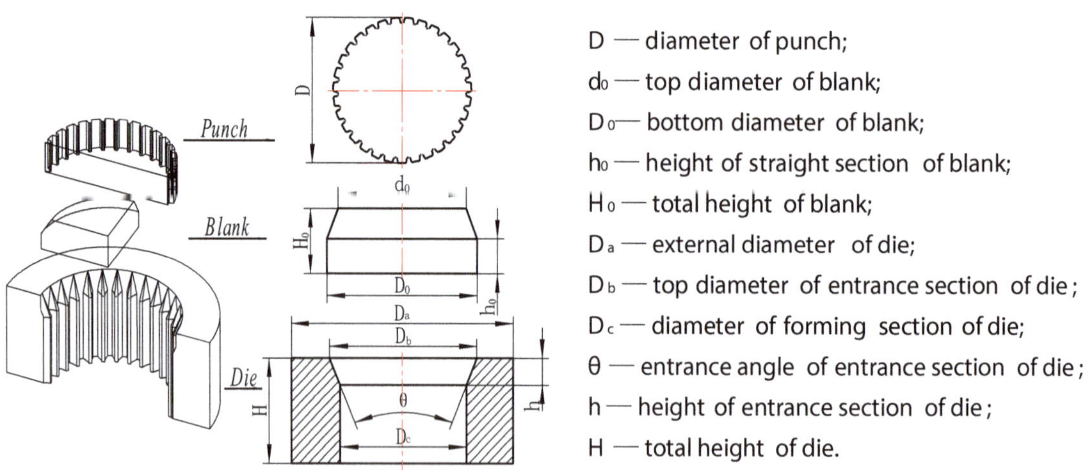

Figure 1. The process model of open-die warm extrusion of spur gear.

forming process, besides the lateral extrusion for gear teeth forming, backward extrusion caused by friction occurs simultaneously, so the initial blank is chamfered in its head.

2.2. Forming principle

The forming principle and load-stroke curve are shown in **Figure 2**. It consists of three stages. In stage OA, the bottom of the blank is divided by the entrance section. In stage AB, the bottom of the blank is formed into gear tooth shape, and the blank is divided continually. In stage BC, the blank entirely gets into the forming section and the gear teeth go through truing until the blank breaks away from the die.

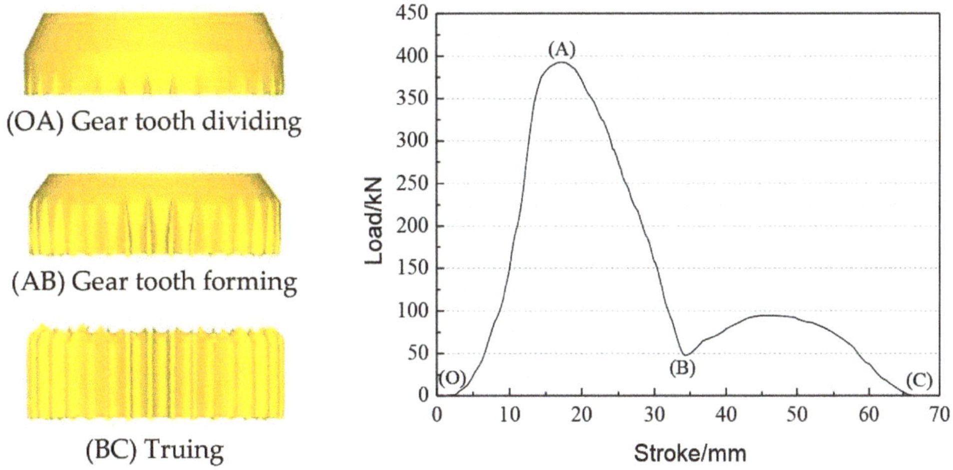

Figure 2. Forming principle and load-stroke curve.

3. Influence factors analysis

Based on DEFORM-3D, a commercial finite element (FE) software with rigid plastic model and shear friction model ($\tau_f = mk$), the FE model of the open-die warm extrusion is established to explore the influence factors on the forming quality. The FE model is shown in **Figure 3**. The

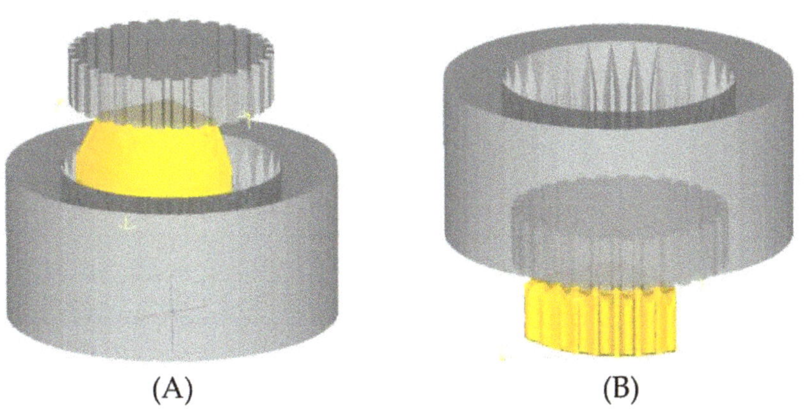

Figure 3. The FE model. (A) Before forming; (B) after forming.

detailed parameters of the FE model are listed in **Table 1**. Eight teeth are chosen in the FE model for better clarity. The element type is tetrahedron, element number is 50,000, node number is 11,249, and minimum edge length is 0.4617 mm.

The foreseeable factors' influence on the forming quality include initial blank size, entrance angle of die, friction factor, extrusion speed, and forming temperature. Four levels are set up for each factor (as listed in **Table 2**) to constitute the $L_{16}(4^5)$ orthogonal array. It should be noted that the initial blank size is determined by the equal volume principle and the height of the straight section (h_0) is determined as a variable. The orthogonal array and experimental results are listed in **Table 3**.

Through the orthogonal experiment, the lubricating condition, entrance angle, and initial blank size are determined as the crucial factors on the forming quality. However, the experimental results in the orthogonal array suggest different forming defects such as insufficient section, material accumulation, and defect section. So, it is unable to make quantitative analysis. Therefore, the quantitative analysis of each crucial factor is investigated by using the single factor experiment.

3.1. Lubricating condition

Lubricating condition is the most significant factor influence on the forming quality. When the friction factor is large, the insufficient section will be formed at the bottom of workpiece, and the flashes will be formed at the top of workpiece as shown in **Figure 4A**. This is due to the metal material backward flows seriously caused by large friction. The influence of the

Parameter	Tooth number	Modules	Tooth thickness	Material	Die temperature	λ_1	λ_2
Value	31	2	20 mm	$20Cr_2Ni_4A$	473.15K	5N/(s·mm·C)	0.02N/(s·mm·C)

λ_1 is the heat transfer coefficient between the blank and the die.
λ_2 is the convection coefficient to environment.

Table 1. Detailed parameters of the FE model.

Factor	Level			
	1	2	3	4
Straight section height h_0 (mm)	5	8	10	15
Entrance angle θ (°)	20	36	52	78
Forming temperature (K)	973.15	1073.15	1173.15	1273.15
Friction factor	0.2	0.3	0.5	0.7
Extrusion speed V (mm·s^{-1})	10	20	30	40

Table 2. Factors and levels.

Scheme	Straight section height h_0 (mm)	Entrance angle θ (°)	Forming temperature (K)	Friction factor	Extrusion speed (mm·s^{-1})	Experimental result
1	5	20	973.15	0.2	10	Defect section
2	5	36	1073.15	0.3	20	Defect section
3	5	52	1173.15	0.5	30	Insufficient section
4	5	78	1273.15	0.7	40	Insufficient section
5	8	20	1073.15	0.5	40	Defect section
6	8	36	973.15	0.7	30	Insufficient section
7	8	52	1273.15	0.2	20	Defect section
8	8	78	1173.15	0.3	10	Material accumulation
9	10	20	1173.15	0.7	20	Insufficient section
10	10	36	1273.15	0.5	10	Insufficient section
11	10	52	973.15	0.3	40	Defect section
12	10	78	1073.15	0.2	30	Material accumulation
13	15	20	1273.15	0.3	30	Insufficient section
14	15	36	1173.15	0.2	40	Insufficient section
15	15	52	1073.15	0.7	10	Material accumulation
16	15	78	973.15	0.5	20	Material accumulation

Table 3. $L_{16}(4^5)$ orthogonal array and experimental results.

lubricating condition on maximum forming load and the height of insufficient section are shown in **Figure 4B**. The friction factor is set up from 0.2 to 0.5 at an interval of 0.1, with the fixed parameters: entrance angle at 36°, height of straight section at 10 mm, forming temperature at 1073.15 K, and extrusion speed at 10 mm·s^{-1}. It indicates a better lubricating condition with a smaller height of insufficient section and a smaller forming load. It should be noted that

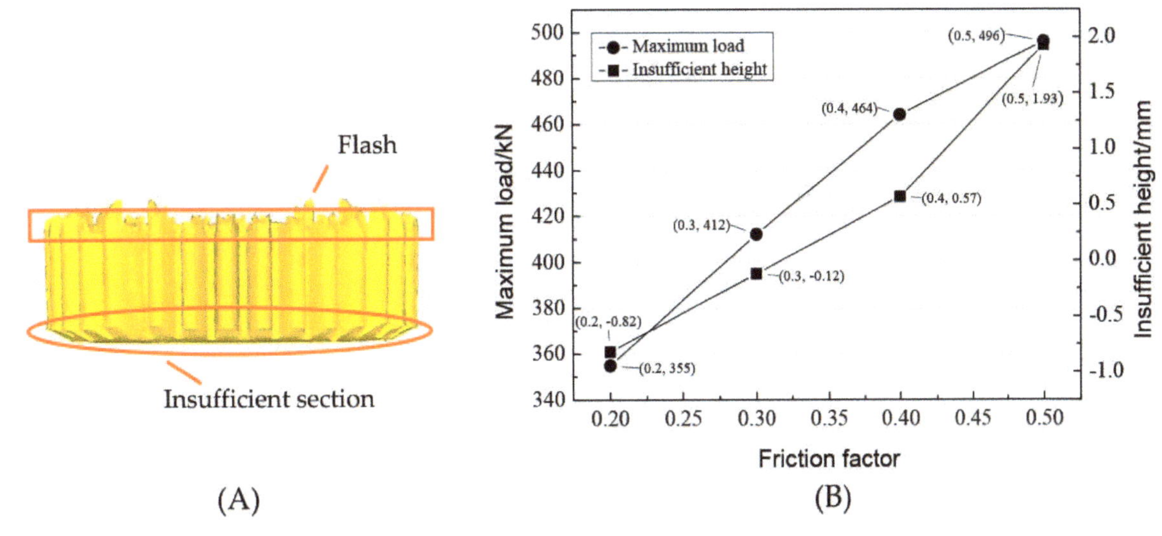

Figure 4. Effects of lubricating condition. (A) Forming result in poor lubricating condition; (B) influence of lubricating condition.

when the friction factor is 0.2, the height of the insufficient section presents a negative value (−0.82 mm). This is because the passageway of die is progressively diminishing. When the lubricating condition is very good, the blank will occur forward extrusion in the gear tooth dividing stage. So the addendum is elongated along the extrusion direction.

3.2. Entrance angle

Entrance angle of die is the second significant factor. The blank cannot be divided sufficiently when the entrance angle is too large, which will lead to the materials accumulation in the gear tooth-forming stage, as shown in **Figure 5A**. The influence of the entrance angle on the height of the defect section in a good lubrication (friction factor 0.3) condition and the height of the insufficient section in a poor lubrication (friction factor 0.7) condition is shown in **Figure 5B**. An exhibition of the defect section is shown in **Figure 7A**. The entrance angle is set up at 20, 36, and 52°, with the fixed parameters: friction factor at 0.3, height of straight section at 10 mm, forming temperature at 1073.15 K, and extrusion speed at 10 mm·s^{-1}. It reveals that both heights of the defect section and the insufficient section decrease with the increase of the entrance angle on the premise of no materials accumulation. But it should be noted that the larger entrance angle leads to the increase of forming load in the gear tooth dividing stage as shown in **Figure 6**. When the entrance angle is 52°, the forming load is markedly increased that is harmful to die life. When the entrance angle is 36°, the forming load of the truing stage is higher than others (the dotted area in **Figure 6**). It means that the contact between the work-piece and the die is better, namely, the gear tooth has a better forming result. So, 36° is a more reasonable choice for the entrance angle.

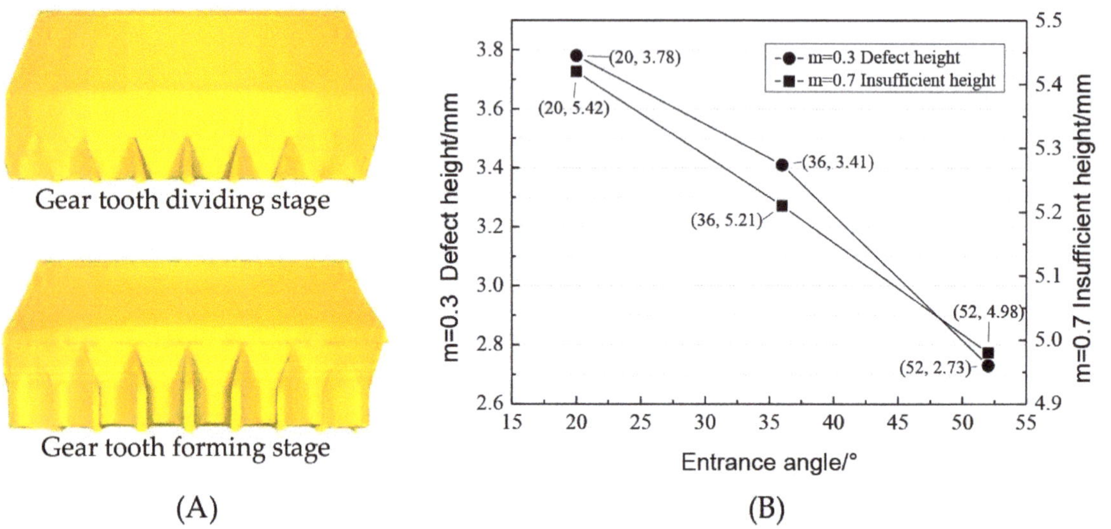

Figure 5. Effects of the entrance angle. (A) Material accumulation when entrance angle is 78°; (B) influence of entrance angle.

3.3. Initial blank size

If process parameters are set up unreasonably, defect sections will be generated at the top of the blank. The defect section will be aggravated if the straight section of the blank is too short (**Figure 7A**). Conversely, if the straight section is too long, the burr will be formed by

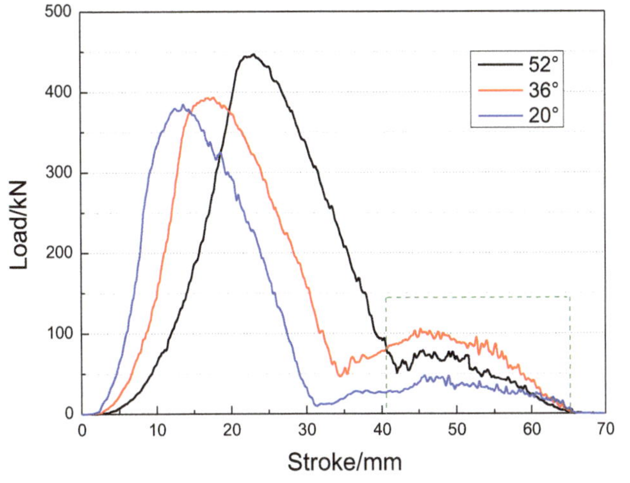

Figure 6. Effects of the entrance angle on forming load.

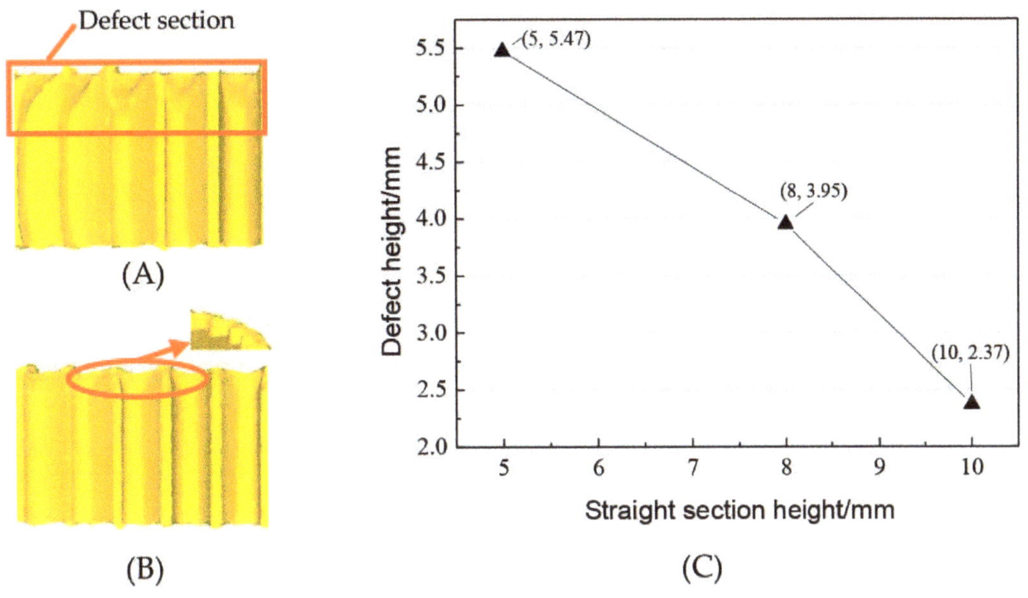

Figure 7. Effects of initial blank size. (A) Straight section 5 mm; (B) Straight section 15 mm; (C) Influence of straight section height.

non-filling materials (**Figure 7B**). The influence of the initial blank size on the height of the defect section is shown in **Figure 7C**. The height of the straight section is set up at 5, 8, 10, and 15 mm, with the fixed parameters: entrance angle at 36°, friction factor at 0.3, forming temperature at 1073.15 K, and extrusion speed at 10 mm·s^{-1}. It is observed that the longer straight section results in the shorter defect section and better forming quality. Even so, the burr should be avoided by selecting a reasonable initial blank size.

3.4. Preliminary experiment

Through the above analysis, the optimized parameters of three crucial factors are ascertained as friction factor 0.3, entrance angle 36°, and straight section 10 mm. The forming temperature is set as 1073.15 K. Because of the temperature fluctuation, the experimental heating temperature is limited to 1048.15 ± 25 K. The reason is that the plasticity of $20Cr_2Ni_4A$ steel is good and the oxidation is not serious in this temperature. According to the parameters of experimental equipment, the extrusion speed is chosen as 10 mm·s^{-1}. Based on the above parameters, the experimental schemes are selected. The experimental equipment is YA-315T hydraulic press and the heating equipment is GP-60 ultra-audio-frequency induction heater. Preheating temperature on the die is 473.15–523.15 K. The experimental material is $20Cr_2Ni_4A$ steel. Oil-based graphite, water-based graphite, and lead oxide are, respectively, chosen as lubricants. **Figure 8** shows the initial blank, the die, and the experimental samples. Unfortunately, all of the experimental schemes do not achieve the desired forming results. The forming effects are similar to the finite element simulation results when the friction factor is 0.6. The insufficient section generates at the bottom of the workpiece, and there still are underfilling teeth.

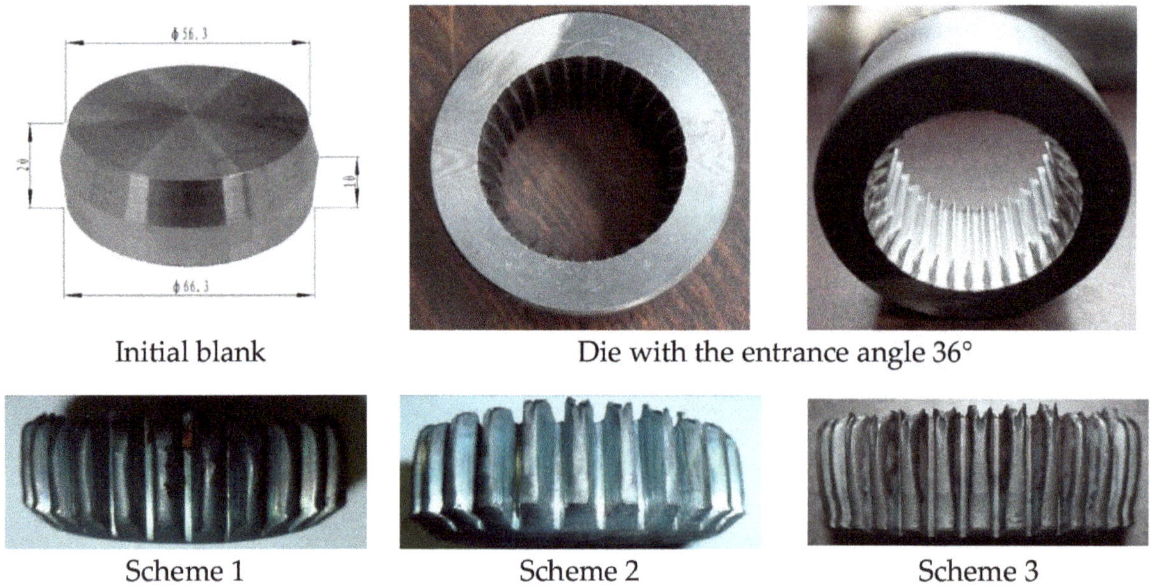

Figure 8. Initial blank, die, and experimental samples.

4. Optimum die design

Through the preliminary experiment studies, it is found that a good lubricating condition is extremely difficult to achieve caused by the complex contact surfaces between the blank and the die. A pursuit of good lubricating condition will certainly put forward rigorous demands on the die surface quality and the selection of lubricants, which will result in an increased production cost. In addition, the lubricating condition will certainly become worse with continued production. So, it is important to study how the good-quality spur gears can also be formed in a poor lubricating condition.

4.1. Defect cause analysis

An important reason for insufficient section is the deterioration of lubricating condition. **Figure 4B** shows the influence of the lubricating condition on the height of insufficient section. The insufficient section almost disappears in good lubricating condition. With the deterioration of the lubricating condition, the height of insufficient section tends to increase. **Figure 9** shows the velocity fields for two cases. One workpiece with insufficient section is shown in **Figure 9A** and another workpiece without defect is shown in **Figure 9B**. It can be seen from the figures that the velocity field is unbalanced in the case of the insufficient section and the generating area is located in the transition zone from entrance section to forming section. After the workpiece passes through the transition zone into the forming section, the velocity field tends to be more balanced. So, the die structure of transition zone is also an important factor that causes the insufficient section.

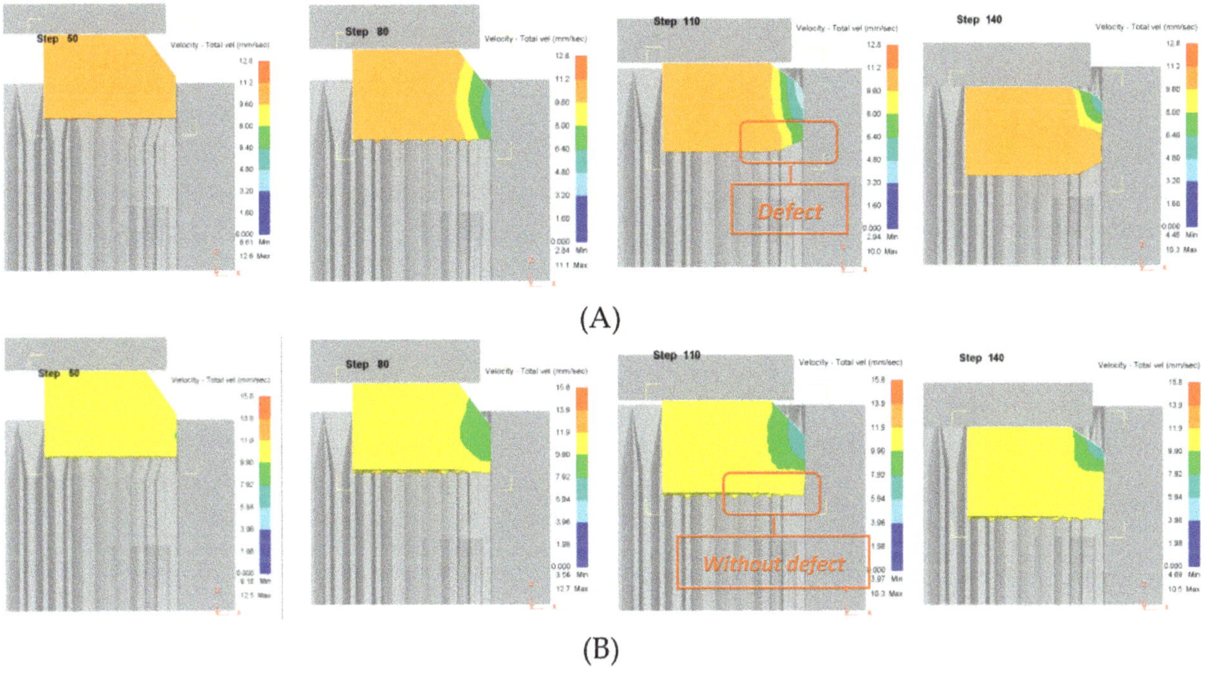

Figure 9. Velocity field of workpiece during open-die warm extrusion. (A) Workpiece with insufficient section; (B) workpiece without insufficient section.

Through the above analysis, the main reason causing the insufficient section of gear face is the unbalanced velocity field during the extrusion process—lubricating condition and die structure being two decisive factors. (1) Lubricating condition: the friction between the workpiece and the die increases with the deterioration of lubricating condition, and the reverse axial strain dominates the deformation results in the insufficient section. (2) Die structure: the bottom of the blank is divided by the entrance section. Then, the material flows into the forming section to form gear tooth. In this step, the passageway is progressively diminishing. If the metal flow velocity is under serious disequilibrium, the hard-deformation zone will be generated, resulting in the insufficient section. So, improvement of the lubricating condition and adjustment of the die structure are two corresponding control strategies. The improvement of the lubricating condition is difficult and costly. Therefore, in this chapter, the method is focused on the optimized design of the die structure for defects control and the improvement of forming quality.

4.2. Design variables

Profile variable ratio at the entrance section (*PVR*):

$$PVR = \frac{S_e - S_f}{S_e} \quad (1)$$

where S_e is the axial projection area on the top of the entrance section and S_f is the axial projection area on the bottom of the entrance section (it is also the axial projection area of the forming section). **Figure 10** shows the sketch of *PVR*, which is associated with the contour offset of forming section (Δ), and it decreases with the increase of Δ, as shown in **Figure 11**. So, *PVR* can be indicated by Δ.

Profile variable rate at the entrance section (PVR'):

$$PVR' = \frac{PVR}{h} \quad (2)$$

where h is the height of the entrance section and is associated with the entrance angle of entrance section (θ), and it decreases with the increase of θ. So, PVR' increases with the increase of θ, and it can be indicated by θ.

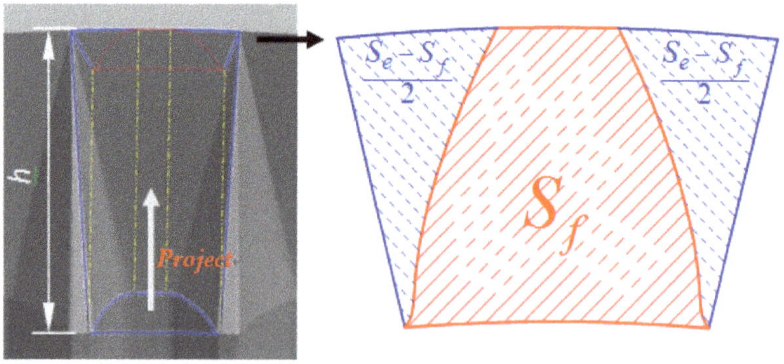

Figure 10. Sketch of *PVR*.

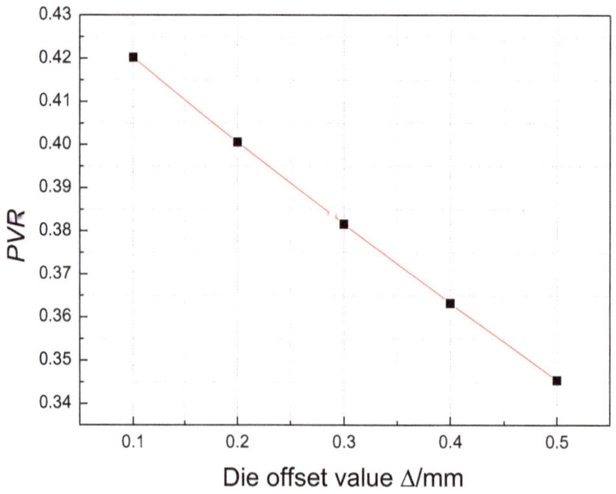

Figure 11. Relationship between *PVR* and Δ.

Extrusion speed (V):

The extrusion speed is also an important influence factor on metal flow velocity. So, V is set as the third design variable. According to the equipment capacity, the extrusion speed ranges from 10 to 50 mm/s.

4.3. Objective function

The evaluation index of forming usually includes forming load, forming quality, die life, etc. The forming load of warm extrusion is much less than forging; the common hot work tool steel can meet all performance requirements for warm extrusion die. So, forming quality is the only important factor in evaluation index for objective functions.

The evaluation index of the height of insufficient section is

$$Min : f_1 = \frac{H_s}{H_g} \tag{3}$$

where H_s is the height of insufficient section and H_g is the total height of workpiece.

The evaluation index of filling quality of gear tooth is

$$Min : f_2 = 1 - \frac{R_g}{R_d} \tag{4}$$

where R_g is the radius of the addendum circle of the formed gear and R_d is the radius of addendum of toothed die. $R_g/R_d = \lambda$ and λ can be called tooth filling coefficient.

4.4. Experimental design and results

This optimized design is a multi-objective optimization problem with three variables and two objectives. The objective functions are optimized by using the response surface method (RSM), and the levels of design variables and process parameters are listed in **Table 4**. The result of the preliminary experiment is similar to the result of numerical simulation when the friction factor is 0.6. So, the friction factor is set as 0.6. Fifteen experimental schemes are established by using central composite design with three factors and three levels. The experimental schemes and results are listed in **Table 5**. The prediction models of f_1 and f_2 are respectively established by using the complete second-order model:

Design variable	Δ (mm) (PVR)	θ (°) (PVR′)	V (mm/s)	Forming temperature (K)	Friction factor
Low level	0.1 (PVR 0.4202)	20 (PVR′ 0.0165)	10	1073.15	0.6
High level	0.5 (PVR 0.3454)	70 (PVR′ 0.0654)	50		
Center level	0.3 (PVR 0.3816)	45 (PVR′ 0.0318)	30		

Table 4. Design variables and process parameters.

Scheme	Δ (mm) (*PVR*)	θ (°) (*PVR′*)	*V* (mm/s)	f_1	f_2
1	0.3 (*PVR* 0.3816)	45 (*PVR′* 0.0351)	50	0.163636	0.003580
2	0.1 (*PVR* 0.4202)	20 (*PVR′* 0.0165)	10	0.433333	0.002039
3	0.1 (*PVR* 0.4202)	20 (*PVR′* 0.0165)	50	0.213333	0.000914
4	0.3 (*PVR* 0.3816)	45 (*PVR′* 0.0351)	30	0.150909	0.002196
5	0.5 (*PVR* 0.3454)	45 (*PVR′* 0.0318)	30	0.047273	0.000897
6	0.3 (*PVR* 0.3816)	70 (*PVR′* 0.0594)	30	0.184848	0.003966
7	0.1 (*PVR* 0.4202)	45 (*PVR′* 0.0387)	30	0.262424	0.001560
8	0.1 (*PVR* 0.4202)	70 (*PVR′* 0.0654)	10	0.456364	0.013768
9	0.5 (*PVR* 0.3454)	20 (*PVR′* 0.0135)	50	0.134182	0.001103
10	0.5 (*PVR* 0.3454)	70 (*PVR′* 0.0537)	50	0.129091	0.003332
11	0.5 (*PVR* 0.3454)	20 (*PVR′* 0.0135)	10	0.111515	0.001101
12	0.1 (*PVR* 0.4202)	70 (*PVR′* 0.0654)	50	0.292727	0.008369
13	0.3 (*PVR* 0.3816)	20 (*PVR′* 0.0150)	30	0.173333	0.000624
14	0.3 (*PVR* 0.3816)	45 (*PVR′* 0.0351)	10	0.229697	0.003399
15	0.5 (*PVR* 0.3454)	70 (*PVR′* 0.0537)	10	0.059394	0.002732

Table 5. Experimental schemes and results.

$$\begin{aligned} f_1 = {} & 0.68869 - 0.91381\Delta - 3.40655 \times 10^{-3}\alpha - 0.014273V \\ & - 3.99091 \times 10^{-3}\Delta\alpha + 0.014875\Delta V + 2.58485 \times 10^{-5}\alpha V \\ & + 0.097980\Delta^2 + 4.50586 \times 10^{-5}\alpha^2 + 1.14343 \times 10^{-4}V^2 \end{aligned} \tag{5}$$

$$\begin{aligned} f_2 = {} & 2.35514 \times 10^{-3} + 6.00054 \times 10^{-3}\Delta + 1.34600 \times 10^{-4}\alpha \\ & - 3.51418 \times 10^{-4}V - 3.83087 \times 10^{-4}\Delta\alpha + 2.22694 \times 10^{-4}\Delta V \\ & - 9.18613 \times 10^{-7}\alpha V - 6.97470 \times 10^{-3}\Delta^2 + 1.26034 \times 10^{-6}\alpha^2 \\ & + 4.95393 \times 10^{-6}V^2 \end{aligned} \tag{6}$$

Significance of each prediction model is analyzed by using analysis of variance; the results are respectively listed in **Tables 6** and **7**. The *R*-squared of f_1 and f_2, respectively, are 0.998569 and 0.930357; the Adj *R*-squared, respectively, are 0.995995 and 0.804999 and the *P*-values respectively are < 0.0001 and 0.0200. These data indicate that the response value is extremely significant. The prediction models both have good accuracy.

4.5. Response surface analysis

The 3D response surface plots and contour plots of f_1 are shown in **Figures 12–14**. The response f_1 decreases with the decrease of *PVR* (increase of Δ); it first decreases and then increases with the increase of *PVR′* (θ); similarly, with the increase of *V*, the response f_1 first decreases and then increases.

Source	Sum of squares	df	Mean square	*F* value	*P*-value
					Prob > *F*
Model	0.198278	9	0.022031	387.805855	< 0.0001
Δ	0.138469	1	0.138469	2437.441605	< 0.0001
α	0.000322	1	0.000322	5.664562	0.0632
V	0.012769	1	0.012769	224.765496	< 0.0001
$\Delta\alpha$	0.003185	1	0.003185	56.073317	0.0007
ΔV	0.028322	1	0.028322	498.547451	< 0.0001
αV	0.001336	1	0.001336	23.522461	0.0047
Δ^2	0.000039	1	0.000039	0.695264	0.4424
α^2	0.002039	1	0.002039	35.898160	0.0019
V^2	0.005379	1	0.005379	94.688992	0.0002
Residual	0.000284	5	0.000057		
Standard deviation: 0.007537, *R*-squared: 0.998569. press: 0.002699, and Adj *R*-squared: 0.995995.					

Table 6. Variance analysis of prediction model f_1.

Source	Sum of squares	df	Mean square	*F* value	*P*-value
					Prob > *F*
Model	0.000158	9	0.000018	7.421595	0.0200
Δ	0.000031	1	0.000031	12.891600	0.0157
α	0.000069	1	0.000069	29.359567	0.0029
V	0.000003	1	0.000003	1.390494	0.2914
$\Delta\alpha$	0.000029	1	0.000029	12.376901	0.0170
ΔV	0.000006	1	0.000006	2.676781	0.1627
αV	0.000002	1	0.000002	0.711676	0.4374
Δ^2	0.000001	1	0.0000001	0.084398	0.7831
α^2	0.000002	1	0.000002	0.672817	0.4494
V^2	0.000010	1	0.000010	4.257755	0.0940
Residual	0.000012	5	0.000002		
Standard deviation: 0.001540, *R*-squared: 0.930357, press: 0.000149, and Adj *R*-squared: 0.804999.					

Table 7. Variance analysis of prediction model f_2.

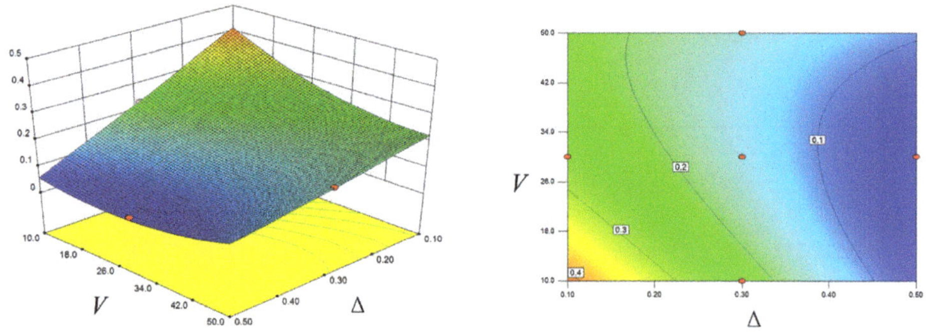

Figure 12. Interaction between V and Δ on f_1 with $\theta = 45°$.

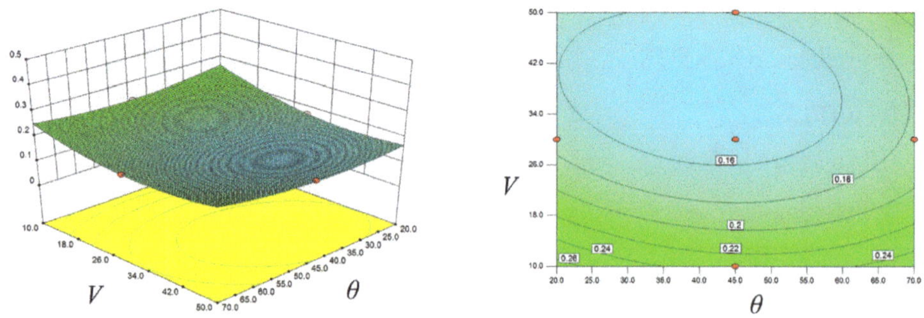

Figure 13. Interaction between V and θ on f_1 with $\Delta = 0.3$ mm.

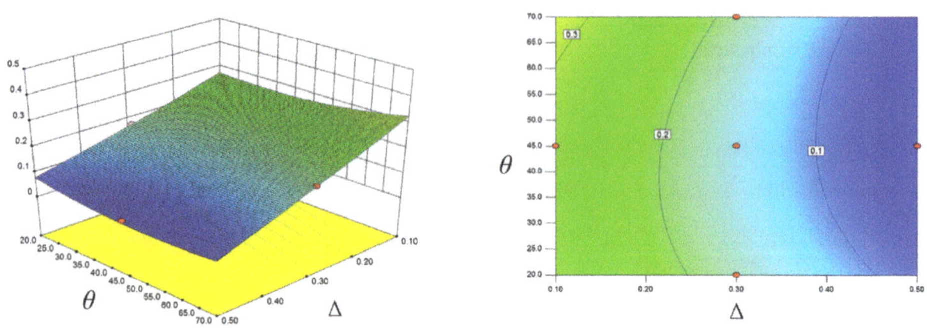

Figure 14. Interaction between θ and Δ on f_1 with $V = 30$ mm/s.

Reason analysis: (1) with the decrease of *PVR* (increase of Δ), the channel of the metal axial flow is enlarged; the variable ratio from the entrance section to the forming section is reduced; the axial flow resistance of the metal is reduced, and so the height of insufficient section is decreased. (2) When the extrusion speed V is slower, the temperature drop of workpiece is quicker, the plasticity of metal is decreased, and then the axial flow of metal is weakened. With the increase of the extrusion speed, the metal deformation energy cannot be timely released, which is conducive to metal flow. However, the extrusion speed continues to increase, the blank does not have enough time for tooth graduation, and this results in the increase of the height of insufficient section. (3) If the entrance angle θ (PVR') is too small, namely, the height of the entrance section (h) is too high, although the workpiece is fully graduated, the temperature of the workpiece is significantly cooled before into the forming section, and then the plasticity of metal is decreased, resulting in the height of higher insufficient section. With the increase of θ, this issue is alleviated; the height of insufficient section is decreased. If the entrance angle (θ) is too large, the workpiece cannot be fully graduated, and the hard-deformation zone will be generated from the material accumulation in the forming stage, and then the height of insufficient section is increased. When at high extrusion speeds, the height of insufficient section is increased with the increase of θ, but not decreased, this due to the temperature drop becomes a secondary factor, and the die structure becomes a determining factor.

The 3D response surface plots and contour plots of f_2 are shown in **Figures 15–17**. The response f_2 decreases with the decrease of PVR (increase of Δ); it increases with the increase of PVR' (θ); it first decreases and then increases with the increase of V.

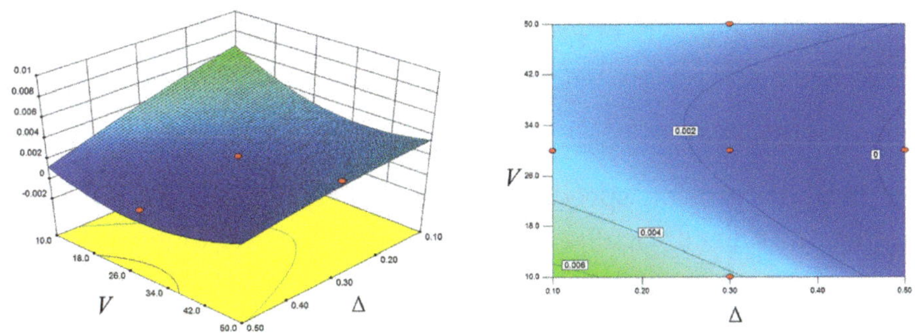

Figure 15. Interaction between V and Δ on f_2 with $\theta = 45°$.

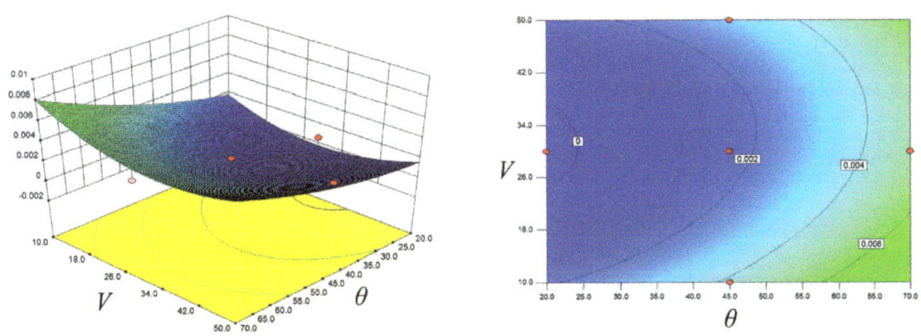

Figure 16. Interaction between V and θ on f_2 with $\Delta = 0.3$ mm.

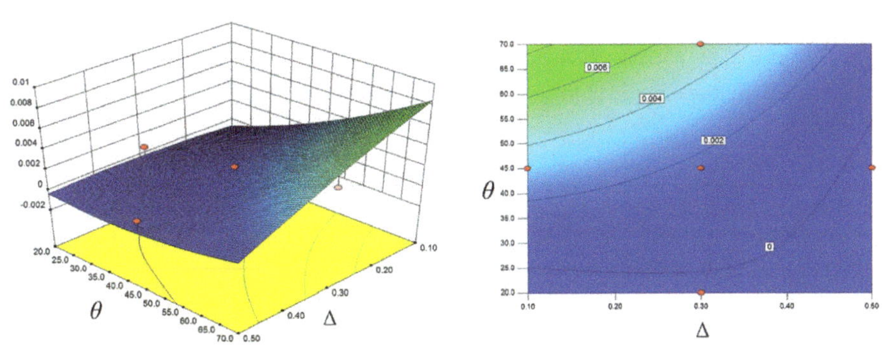

Figure 17. Interaction between θ and Δ on f_2 with $V = 30$ mm/s.

Reason analysis: (1) with the decrease of *PVR* (increase of Δ), the channel of metal radial flow is enlarged; the radial flow resistance of the metal is reduced, and so filling quality of gear tooth is improved. (2) When the extrusion speed *V* is slower, the temperature drop of workpiece is quicker. The plasticity of metal is decreased, and then the radial flow of metal is weakened. With the increase of the extrusion speed, the metal deformation energy cannot be timely released, which is conducive to metal radial flow. However, the extrusion speed continues to increase; the radial flow is weakened because the rigid translation of inner zone and axial deformation of outer zone become predominant, so the filling quality of gear tooth becomes poor. This rule is same with the influence of *V* on response f_1. When the extrusion speed is too high, the height of insufficient section is higher and the filling quality of gear tooth is worse. (3) With the increase of PVR' (θ), the height of entrance section (h) is decreased. The ability of graduation for the workpiece by the entrance section is decreased; if the workpiece cannot be fully graduated, the filling quality of gear tooth is worse.

4.6. Experimental verification

The die structure and extrusion speed are optimized by using the RSM with the goals of decreasing the height of insufficient section and improving the filling quality of gear tooth. The optimal values for each factor are $\Delta = 0.49$ mm ($PVR = 0.3471$), $\theta = 46.02°$ ($PVR' = 0.0327$), and $V = 29.51$ mm/s. The corresponding response values are $f_1 = 0.043854$ and $f_2 = 0.000192$. The theoretical machining allowance of tooth tip, tooth flank, and gear face, respectively, are $\Delta\text{-}R_d f_2$ (0.477 mm), Δ (0.490 mm), and $H_g f_1$ (0.877 mm).

The numerical simulations are done by adopting the original and the optimized parameters, respectively. The results are shown in **Figure 18**. The response values and the relative error of the simulated results with the optimal results are listed in **Table 8**. It is observed that the

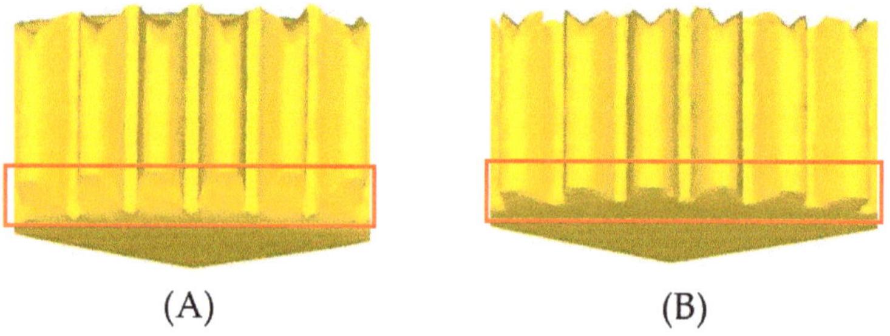

Figure 18. Numerical simulation results. (A) Original parameters; (B) optimized parameters.

Response value	Simulated results with original parameters	Optimized results	Simulated results with optimized parameters	Relative error between simulated and optimized results
f_1	0.237006	0.043854	0.043939	0.1938%
f_2	0.001186	0.000192	0.000189	1.5625%

Table 8. Response values and relative error.

desired optimal result is obtained to decrease the height of insufficient section and improve the filling quality of gear tooth.

The experimental sample is shown in **Figure 19**. The machining allowance is tested by using the 3D coordinate measuring machine. The measuring points of each tooth tip and flank are equispaced at 25 points. As for gear face, each location is measured five times to calculate the mean value. The measurement results are shown in **Figure 20**. Due to the elastic deformation of die and the elastic recovery of workpiece, all the measured values of the tooth flank are with positive deviation.

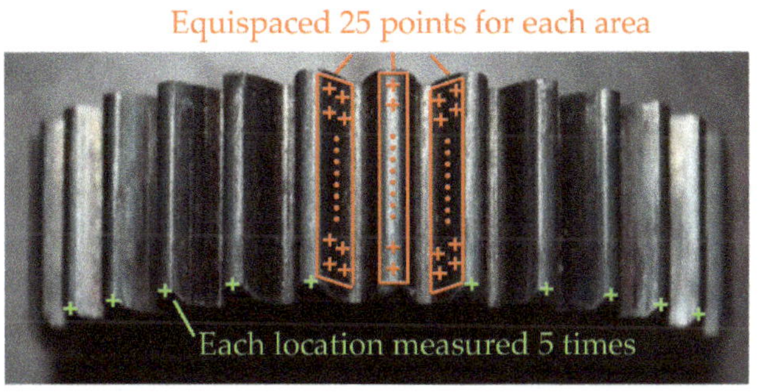

Figure 19. Experimental sample.

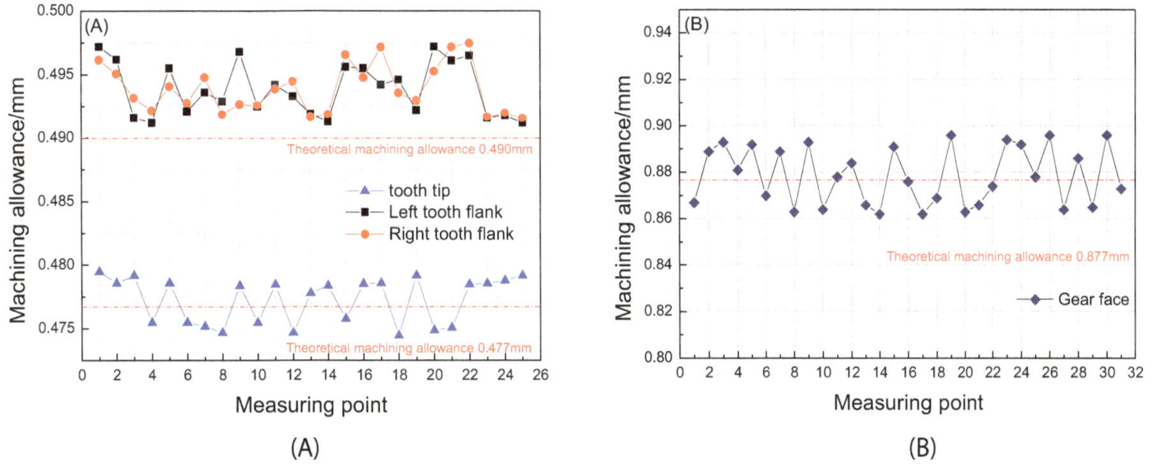

Figure 20. Machining allowance measurement. (A) Measurement result of tip and flank; (B) measurement result of gear face.

5. Variable contour two-step warm extrusion

Through the die optimum design, the goals of decreasing the height of insufficient section and improving the filling quality of gear tooth are achieved. But the machining allowance of tooth flank rises from the original design of the 0.30 to 0.49 mm. In order to further reduce the machining allowance, "Variable Contour Two-Step Warm Extrusion" is presented. First, pre-extrusion is implemented for a die in which the contour of the forming section is enlarged from the final-toothed contour. The formed spur gear is allowed to have insufficient section after pre-extrusion. Then, the workpiece is turned around to implement the finish-extrusion by a die with the final-toothed contour for insufficient section compensation and gear tooth sizing. Because the gear tooth has been formed by pre-extrusion, the forming load of finish-extrusion is very small, so the workpiece need not be reheated. **Figure 21A** shows the influence of contour variation on the height of insufficient section when the friction factor is 0.6. It indicates that the height of insufficient section decreases with the increase of contour variation. The insufficient section almost disappears when the contour variation is 0.3 mm. **Figure 21B** shows the simulated results when the contour variation is 0.3 mm.

Figure 22 shows the experimental sample. The accuracy is tested by using the coordinate measuring machine. Theoretical Computer Aided Design (CAD) model is set as a reference. Twenty-five uniformly distributed spaced points are measured for each gear tooth. The mean error is calculated by $\bar{\delta} = \sum_{i=1}^{25} |\delta_i|/25$. The testing result is shown in **Figure 23A**, and the maximum mean error of single tooth is 15.5 μm. An example of the accuracy testing on one tooth is shown in **Figure 23B**, and the maximum absolute error is 42.7 μm.

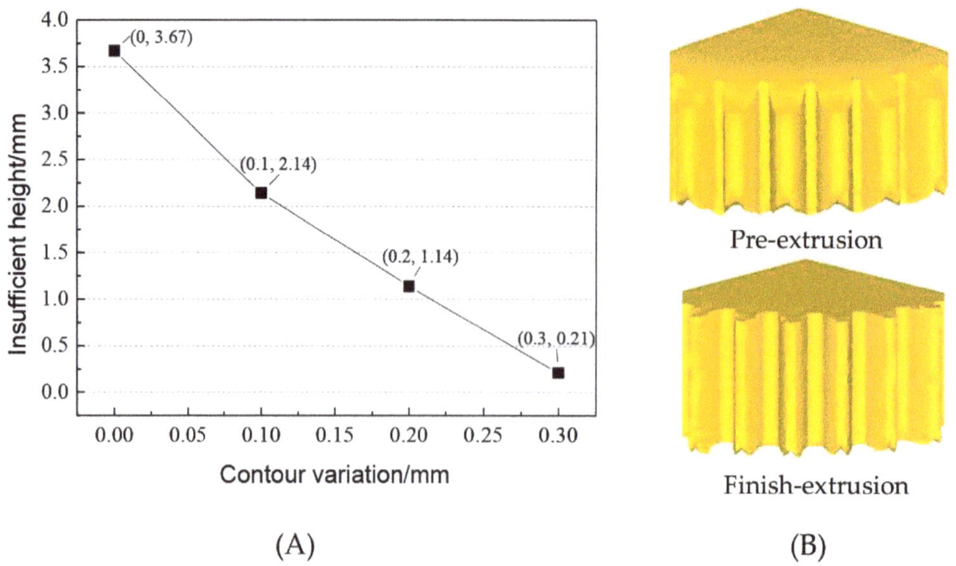

Figure 21. Variable contour two-step warm extrusion. (A) Influence of contour variation on forming quality; (B) simulation results.

Figure 22. Experimental sample.

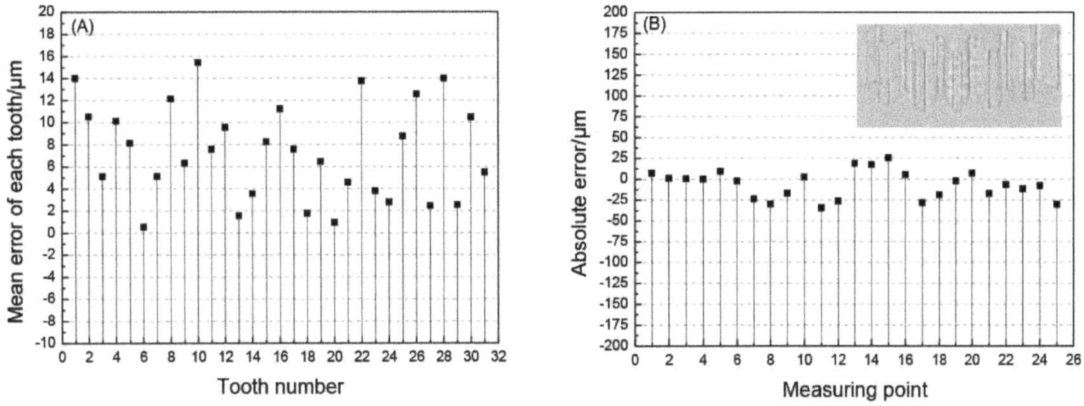

(A) (B)

Figure 23. Testing results. (A) Mean error of each tooth; (B) an example of accuracy testing.

6. Summary

Open-die warm extrusion is developed to replace the traditional manufacturing method of the spur gear. For this technique, lubricating condition, entrance angle, and initial blank size are the crucial factors. Insufficient section will be generated when the lubricating condition is poor. If the entrance angle is too large, material accumulation will occur during the gear tooth forming stage. The defect section will be worse if the straight section is too short.

The main reason for the insufficient section is the unbalanced velocity field. Except for the lubricating condition, the die structure is another decisive factor. The height of insufficient section is decreased with the increase of contour offset. It first decreases and then increases, with the increase of entrance angle. Similarly, with the increase of extrusion speed, it first decreases and then increases. The filling quality of gear tooth improves with the increase of contour offset; it deteriorates with the increase of entrance angle; it first improves and then deteriorates with the increase of extrusion speed.

In order to obtain a good forming result in a poor lubrication, "Variable Contour Two-Step Warm Extrusion" is proposed. Pre-extrusion is for gear teeth forming and finish-extrusion is for insufficient section compensation and tooth sizing. The height of insufficient section decreases with the increase in contour variation. When the contour variation is 0.3 mm, the insufficient section is almost eliminated.

Author details

Wei Wang and Jun Zhao*

*Address all correspondence to: zhaojun@ysu.edu.cn

Key Laboratory of Advanced Forging & Stamping Technology and Science, Yanshan University, Ministry of Education of China, Qinhuangdao, PR China

References

[1] Dean TA. The net-shape forming of gears. Materials and Design. 2000;**21**:271-278

[2] Abdul NA, Dean TA. An analysis of the forging of spur gear forms. International Journal of Machine Tool Design and Research. 1986;**26**:113-123

[3] Chitkara NR, Bhutta MA. Near-net shape forging of spur gear forms: an analysis and some experiments. International Journal of Mechanical Sciences. 1996;**38**:891-916

[4] Tuncer C, Dean TA. Die design alternatives for precision forging hollow parts. International Journal of Machine Tools and Manufacture. 1987;**27**:65-76

[5] Tuncer C, Dean TA. Precision forging hollow parts in novel dies. Journal of Mechanical Working Technology. 1988;**16**:39-50

[6] Cai J, Dean TA, Hu ZM. Alternative die designs in net-shape forging of gears. Journal of Materials Processing Technology. 2004;**150**:48-55

[7] Kondo K, Ohga K. Precision cold die forging of a ring gear by divided flow method. International Journal of Machine Tools and Manufacture. 1995;**35**:1105-1113

[8] Choi JC, Choi Y. Precision forging of spur gears with inside relief. International Journal of Machine Tools and Manufacture. 1999;**39**:1575-1588

[9] Wang GC, Zhao GQ, Xia SS, Luan YG. Numerical and experimental study on new cold precision forging technique of spur gears. Transactions of Nonferrous Metals Society of China. 2003;**13**:798-802

[10] Behrens BA, Odening D. Process and tool design for precision forging of geared components. International Journal of Material Forming. 2009;**2**:125-128

[11] Hu CL, Wang KS, Liu QK. Study on a new technological scheme for cold forging of spur gears. Journal of Materials Processing Technology. 2007;**187-188**:600-603

[12] Zuo B, Wang BY, Li Z, Zheng MN, Zhu XX. Design of relief-cavity in closed-precision forging of gears. Journal of Central South University. 2015;**22**:1287-1297

[13] Song JH, Im YT. Determination of a major design parameter for forward extrusion of spur gears. Journal of Manufacturing Science and Engineering. 2004;**126**:255-263

[14] Song JH, Im YT. The applicability of process design system for forward extrusion of spur gears. Journal of Materials Processing Technology. 2007;**184**:411-419

[15] Yuan AF. Cold extrusion of a long trapezium spline and its forming analysis. International Journal of Advanced Manufacturing Technology. 2009;**41**:461-467

[16] Jeong MS, Lee SK, Yun JH, Sung JH, Kim DH, Lee S, Choi TH. Green manufacturing process for helical pinion gear using cold extrusion process. International Journal of Precision Engineering and Manufacturing. 2013;**14**:1007-1011

[17] Niechajowicz A, Tobota A. Warm deformation of carbon steel. Journal of Materials Processing Technology. 2000;**106**:123-130

[18] Li YY, Zhao SD, Fan SQ, Yan GH. Study on the material characteristic and process parameters of the open-die warm extrusion process of spline shaft with 42CrMo steel. Journal of Alloys and Compounds. 2013;**571**:12-20

[19] Wang W, Zhao J, Zhai RX. A forming technology of spur gear by warm extrusion and the defects control. Journal of Manufacturing Processes. 2016;**21**:30-38

[20] Wang W, Zhao J, Zhai RX, Ma R. Variable contour two-step warm extrusion forming of spur gear and the deformation behavior of $20Cr_2Ni_4A$ steel. International Journal of Advanced Manufacturing Technology. 2017;**88**:3163-3173

Permissions

All chapters in this book were first published in EMPFP, by InTech Open; hereby published with permission under the Creative Commons Attribution License or equivalent. Every chapter published in this book has been scrutinized by our experts. Their significance has been extensively debated. The topics covered herein carry significant findings which will fuel the growth of the discipline. They may even be implemented as practical applications or may be referred to as a beginning point for another development.

The contributors of this book come from diverse backgrounds, making this book a truly international effort. This book will bring forth new frontiers with its revolutionizing research information and detailed analysis of the nascent developments around the world.

We would like to thank all the contributing authors for lending their expertise to make the book truly unique. They have played a crucial role in the development of this book. Without their invaluable contributions this book wouldn't have been possible. They have made vital efforts to compile up to date information on the varied aspects of this subject to make this book a valuable addition to the collection of many professionals and students.

This book was conceptualized with the vision of imparting up-to-date information and advanced data in this field. To ensure the same, a matchless editorial board was set up. Every individual on the board went through rigorous rounds of assessment to prove their worth. After which they invested a large part of their time researching and compiling the most relevant data for our readers.

The editorial board has been involved in producing this book since its inception. They have spent rigorous hours researching and exploring the diverse topics which have resulted in the successful publishing of this book. They have passed on their knowledge of decades through this book. To expedite this challenging task, the publisher supported the team at every step. A small team of assistant editors was also appointed to further simplify the editing procedure and attain best results for the readers.

Apart from the editorial board, the designing team has also invested a significant amount of their time in understanding the subject and creating the most relevant covers. They scrutinized every image to scout for the most suitable representation of the subject and create an appropriate cover for the book.

The publishing team has been an ardent support to the editorial, designing and production team. Their endless efforts to recruit the best for this project, has resulted in the accomplishment of this book. They are a veteran in the field of academics and their pool of knowledge is as vast as their experience in printing. Their expertise and guidance has proved useful at every step. Their uncompromising quality standards have made this book an exceptional effort. Their encouragement from time to time has been an inspiration for everyone.

The publisher and the editorial board hope that this book will prove to be a valuable piece of knowledge for researchers, students, practitioners and scholars across the globe.

List of Contributors

Tiwari Ajita
Department of Agricultural Engineering, Assam University, Silchar, India

Carlos Martín Enríquez Castro
Food Science and Technology Program, University of Sonora, Hermosillo, Sonora, Mexico

Patricia Isabel Torres-Chávez, Benjamín Ramírez-Wong, Ana Irene Ledezma-Osuna and María Irene Silvas-García
University of Sonora, Hermosillo, Sonora, Mexico

Armando Quintero-Ramos
Autonomous University of Chihuahua, Chihuahua, Mexico

Jaime López-Cervantes
Technological Institute of Sonora, Obregón, Sonora, Mexico

G.N. Kouzilos, G.V. Seretis, C.G. Provatidis and D.E. Manolakos
National Technical University of Athens, School of Mechanical Engineering, Athens, Greece

Cuauhtémoc Reyes Moreno, Perla C. Reyes Fernández, Edith O. Cuevas Rodríguez, Jorge Milán Carrillo and Saraid Mora Rochín
Faculty of Chemical and Biological Sciences, Autonomous University of Sinaloa, Mexico

Sayyad Zahid Qamar
Mechanical and Industrial Engineering Department, Sultan Qaboos University, Al-Khodh, Muscat, Sultanate of Oman

Martha G. Ruiz-Gutiérrez, Miguel Á. Sánchez-Madrigal and Armando Quintero-Ramos
Faculty of Chemical Sciences, Autonomous University of Chihuahua, Nuevo Campus Universitario Circuito Universitario s/n, Chihuahua, Chihuahua, México

Tasneem Pervez, Sayyad Z. Qamar, Omar S.A. Al-Abri and Rashid Khan
Mechanical and Industrial Engineering Department, College of Engineering, Sultan Qaboos University, Oman

Ivan Kurunov
Novolipetsk Steel (NLMK), Lipetsk, Russia

Aitber Bizhanov
J. C. Steele & Sons, Inc., Statesville, NC, USA

Efren Delgado and Damian Reyes-Jaquez
Food Science and Technology, Department of Family and Consumer Sciences, New Mexico State University, New Mexico, USA

Haichen Zhang
School of Materials Science & Energy Engineering, Foshan University, Foshan, Guangdong, P.R. China

National Engineering Research Center of Novel Equipment for Polymer Processing, South China University of Technology, Guangzhou, Guangdong, P.R. China

Yong Liang
Changzhou Institute of Technology, Changzhou, Jiangsu, P.R. China

Wei Wang and Jun Zhao
Key Laboratory of Advanced Forging & Stamping Technology and Science, Yanshan University, Ministry of Education of China, Qinhuangdao, PR China

Index

A

Agglomeration, 75, 124-126, 131, 142-143

Agro-industrial By-products, 145-147

Alkaline Cooking, 18-20, 23, 25, 30-31

Annealing, 17, 29, 36, 170

Anthocyanin, 9, 67, 71, 83-84, 96-98

Antioxidant Activity, 9, 11, 52, 56-57, 61-67, 70-71, 82-83, 89, 93-94, 98, 100, 158

Aquaculture, 10-11, 78, 145-153, 155-159, 161-163

B

Betaxanthins, 82, 92, 95-96

Bioactive Compounds, 50-53, 56, 61-63, 65, 70-71, 81-85, 89-94, 96-97, 145, 147, 158, 163

Bioavailability, 13, 56, 58, 60-61, 64, 70

Briquetting, 75, 124-128, 142-143

C

Cardiovascular Disease, 52, 59, 63

Cereal Grains, 21, 23, 51-53, 60-61, 63-65, 70

Contact Stress, 103, 117-118

Cottonseed Meal, 145-147, 158-160, 162

Crystallinity, 17-18, 20-21, 148, 154, 166-168, 170

D

Decarboxylation, 62, 83, 89

Deformation, 24, 72, 102-103, 105, 107-109, 113-115, 117, 119-120, 122, 127, 172-173, 177, 187, 192-195, 198-199

Denaturation, 3, 7, 22, 50, 55-58, 85, 148

Die Pressure, 36, 38, 54

Dietary Fiber, 15, 22, 50, 52-53, 59-60, 70, 81-82, 98, 100

Direct Expanded Product, 2

E

Entrance Angle, 179-180, 182-185, 188, 192-193, 197

Enzyme Inhibitors, 7-8, 55, 58

Expanded Snacks, 1-2, 9, 11-12, 98-99

Expansion Ratio, 4-5, 103-105, 110-112, 114, 116-120

Extrudate Expansion, 4-5

Extruded Products, 1-2, 4, 6-7, 10, 12, 16, 21, 31, 36, 50, 55, 57, 60, 62, 66, 73, 81, 84, 87-89, 93, 95, 148-150, 159-160

Extrusion Cooking, 1-4, 8-14, 16, 21, 23, 27, 31-33, 50-51, 54-57, 60-64, 66, 68-70, 77, 80-81, 83-85, 87-89, 93-97, 99-100, 162

Extrusion Moisture, 126, 148-153, 157-158

F

Fatty Acid, 53, 67, 148

Feed Moisture, 4-6, 8-9, 11, 15-16, 21, 25, 56, 59-60, 85, 87, 95

Ferroalloy Production, 124, 139-140, 142-143

Finite Element Analysis, 35, 40-44, 47-48, 109, 111, 121

Finite Element Method, 101-102, 104-105, 109

Fishmeal, 145-153, 155-158, 163

Flow Path, 38, 172, 174

Functional Food, 77, 80-81, 84, 88

G

Gastrointestinal Tract, 56, 61

Gelatinization, 2-3, 5-7, 12, 16-21, 24, 27-28, 31-32, 50, 55-56, 59, 85-87, 154

Gelatinization of Starch, 3, 7, 19, 50

H

Heat Treatment, 60, 71, 73, 132

High-density Polyethylene, 35-38, 40

I

Indirect Extrusion, 74, 101-103

Industrial Trial, 140, 142

L

Legumes, 2, 8, 14, 81, 145-147

Lipid Oxidation, 9, 22, 57, 68

Lubricating Condition, 179, 182-184, 186-187, 197

M

Maillard Reaction, 56-58, 60, 62, 67-69

Material Accumulation, 182-184, 193, 197

Material Velocity, 102, 113

Metallurgy, 73, 124, 142-143

Microorganisms, 9, 50, 56

N

Nixtamalization, 15-16, 18-21, 25-27, 29, 66, 77-78, 80, 83, 153

Nutritional Properties, 1, 22, 32, 50-51, 56-57, 63

Nutritional Value, 7, 13, 22, 25-27, 57, 69, 78, 162

O

Objective Functions, 36, 189

Oil Extraction, 146-147

Open Die, 179

P

Phenolic Compounds, 8, 52-53, 56, 61-65, 67-68, 71, 81, 84, 90, 100

Phytates, 14, 51, 55, 60

Phytochemicals, 9, 51-53, 61-62, 64-66, 71

Polymer Extrusion, 35, 41, 48-49, 75-76, 80

Protein Concentration, 146-147, 158

Q

Quality Parameter, 20, 24, 150, 155

S

Screw Speed, 4-10, 12, 21, 37-38, 55-56, 59, 81, 84-87, 90-95, 100, 153, 162

Single Screw Extruder, 1, 10, 35, 158-159, 171, 178

Specific Mechanical Energy, 6, 86, 153, 162

Spur Gear, 74, 179-180, 195, 197-199

Staple Food, 15, 52, 146

Starch Content, 4, 7, 27, 59, 69, 76, 147, 153-154, 159

Starch Granule, 16-18, 20-23

Stiff Extrusion, 75, 125-126, 135, 138-139, 142

Stiff Vacuum Extrusion, 75, 124-128, 130, 135, 139, 142-144

T

Thermal Treatment, 21-22, 153

Thermoplastic Polymer, 164, 166, 170

Tubular Expansion, 74, 79, 101-107, 109-111, 114, 117, 119, 121-123

U

Ultra-high Molecular Weight Polyethylene, 164, 176-177

V

Viscosity, 5, 7, 17, 20, 22-23, 30, 32, 35, 37, 41, 55, 85-86, 128, 137-138, 165, 168-171, 174, 178

Viscous Dissipation, 35, 41

W

Warm Extrusion, 179-181, 187, 189, 195-197, 199

Water Absorption Index, 6, 86-87, 148-151, 153, 159

CPSIA information can be obtained
at www.ICGtesting.com
Printed in the USA
LVHW062019100122
708212LV00002B/342

9 781639 891955